Black Literature, Social Justice, and the Environment

Diana Leong

Black Literature, Social Justice, and the Environment

Toward a Slave Ship Ecology

palgrave
macmillan

Diana Leong
English and Comparative Literature
San Diego State University
San Diego, CA, USA

ISBN 978-3-032-19287-5 ISBN 978-3-032-19288-2 (eBook)
https://doi.org/10.1007/978-3-032-19288-2

Cover credit: PM-Aerial-03-HC-A5

This Palgrave Macmillan imprint is published by the registered company Springer Nature Switzerland AG
The registered company address is: Gewerbestrasse 11, 6330 Cham, Switzerland

To Paulette, Wendell, and Michele Leong for their love and support

Competing Interests The author has no competing Interests to declare that are relevant to the content of this manuscript.

Contents

List of Figures

CHAPTER 1

Introduction: Against Wind and Tide

An unprecedented increase in the scope, scale, and frequency of ecological disaster in the last generation has intensified public interest in conditions of environmental precarity, and the social inequalities they throw into stark relief. These matters of global concern have forced sustained consideration of how we might address the ecological consequences of industrial activity while attending to the uneven distribution of environmental risks and resources. In response, environmental justice scholars and activists have pointed out that because categories of social identity organize our engagements with nature into hierarchical differences, any attempt to address environmental precarity must simultaneously grapple with the question of what it means to be "free" in an environmentally just world. *Toward a Slave Ship Ecology* argues that situating environmental justice within the longer tradition of international abolitionism will highlight its commitment to an expansive vision of freedom, justice, and equality.

Across the humanities and social sciences, environmental justice is generally described as a twentieth-century social movement that originated with black civil rights activists and their demands for equal protection under the law. While the strategies of environmental justice are certainly indebted to the modern civil rights movement, this historical periodization has produced two effects. First, it encourages the perception of environmental justice as a "special interest group" that

D. Leong, *Black Literature, Social Justice, and the Environment*,
https://doi.org/10.1007/978-3-032-19288-2_1

prioritizes identity politics (i.e., race) over the supposedly broader goals of mainstream environmentalism (e.g., climate change, petrocultures, biodiversity loss). And second, it casts environmental racism as a relatively recent phenomenon precipitated by the failure of state policy and practice since the New Deal. Consequently, environmental justice scholars and activists have increasingly questioned whether a focus on environmental racism is adequate for addressing global environmental issues. Within this "transnational turn," or what ecocritic Ursula Heise identifies as "eco-cosmopolitanism," inquiries about race in general, and antiblackness in particular, are often depicted as narrow (i.e., specific to the United States) or inelastic. In these visions of an environmentally just world, "freedom" appears post-racial. This assessment not only misinterprets the difficulty of developing new approaches to race as a weakness inherent to race-based analyses as such, but also disconnects environmental justice from its theoretical and practical antecedents in the international abolitionist movement.

Such misinterpretations are not new. Like environmental justice, abolitionism emerged out of black political struggles for race, class, and gender equality. However, as historian Manisha Sinha observes in *The Slave's Cause*, abolitionism is frequently reduced to an antebellum movement dominated by white middle-class reformers and "burdened by racial paternalism," a view that obscures the black radicalism and self-emancipatory activities at the heart of the resistance to slavery.[1] Abolitionism was not only a campaign to abolish the institution of slavery in the historic instance, it is also an ongoing project aimed at the destruction of every condition that makes slavery possible in the first place. These conditions include the exploitation of the material environment. As feminist science studies scholar Donna Haraway notes, the "devastating transformation of diverse kinds of human-tended farms, pastures, and forests into extractive and enclosed plantations" is just one example of how modern racial slavery was sustained through ecological means.[2] These same strategies of extraction and enclosure also enable the mechanisms of environmental racism that restrict genuinely democratic participation in the institutions of state and civil society, well into the present day. As such, the race, class, and gender equality pursued by radical abolitionism is both the precondition for and the outcome of any attempt to create an ethical collective inhabitation of the earth. Environmental justice is the ecological imperative of abolitionism.

To probe this connection further, I pair twentieth- and twenty-first-century works of black literature with a range of multimedia texts from the environmental justice and abolitionist movements, paying close attention to how each text constructs and frames its responses to environmental precarity. Because slavery established and complicated the distinctions between humans and nonhumans, and between culture and nature, black writers have consistently formulated critical concepts of both "nature" and "freedom." I therefore read black literature as an environmental justice philosophy oriented toward the wider goals of abolitionism. This methodology seeks to clarify how the historical legacies of slavery have shaped the ways we define and pursue environmental and social well-being. Although literary, feminist, and critical race studies scholars stress that environmental justice is as much about ideology, representation, and narrative as it is about the environment, environmental justice studies itself remain preoccupied with the practical politics of law and policy. Accordingly, I adopt an expansive view of environmental justice that emphasizes its more speculative, aesthetic features. By engaging texts that are both explicitly and implicitly identified with environmental justice, we can better understand how the imaginative registers of the movement impact its overall development.

In recognition of slavery's essential role in these relationships, I present the slave ship as a unique site for the convergence of abolitionism and environmental justice. The book is organized around the ecology of the slave ship as a heuristic for understanding environmental precarity across modernity's *longue durée*. As a complex seaborne ecology, the slave ship represents a matrix of environmental relations and resource regulations that produce ecological models for making and managing racial difference. While both sailors and slaves lived in common conditions of precarity, to mark themselves off as fundamentally different from the enslaved, sailors systematically manipulated access to natural resources and exposure to natural elements to increase the suffering of slaves while ensuring their own survival, or at least their own positional authority. Consequently, the differential ability to access and control such resources, or to be sheltered from such elements, became symbols of racial superiority and inferiority; over time and as an effect of a persistent process of mystification, it appeared as though white bodies were "naturally" robust and black bodies "naturally" deficient. It is in this general context that I take up abolitionist William Lloyd Garrison's 1831 comment that African Americans must "[struggle] against wind and tide…for the advancement of [their] rights."[3] The book suggests, then, that our current ecological predicaments can be conceived of as the global expansion of slave ship ecology and the necessary critical response as an extension of abolitionism or the "slave's cause."

While the term "ecology" traverses a wide range of meanings and references, I am interested specifically in what holds these together as it pertains to the triangulation of race, nature, and slavery. In its scientific sense, ecology was originally conceived of as the study of relations between organisms and their environment, with related fields like evolutionary ecology and behavioral ecology focusing on adaptation, population distribution, succession dynamics, and the circulation of energy and matter. In its colloquial usage, ecology can refer to a network of relations, an ecosystem, or the environment more broadly. And when deployed from within the humanities and social sciences, the term denotes a philosophy of nature (e.g., queer ecology), a concept-metaphor (e.g., digital ecologies), or a framework of analysis (e.g., political ecology). These modalities are intimately entangled, and I mention them here to qualify my own use of the term. One of the biggest challenges for a project focused on ecological matters is the fact that relations between entities and their environments are rarely self-evident. Because these relations must be identified and mapped, designating a set of relations/entities as an ecology necessarily involves practices of meaning-making that unfold on multiple scales. More importantly, these practices are not immune to the libidinal economy of race or the historically given laws of language which are "preselected and limited, by fashion, by cultural norms—by systems that shape us such as gender and race—by what's acceptable."[4]

Research in the environmental humanities, for example, tends to favor process, context, interdependence, and entanglement over the discrete materiality espoused by mechanistic theories of nature. This approach seeks to decenter the human by attending to the complex networks of more-than-human objects, matter, and forces in which the human is situated (e.g., blue humanities, posthumanism, new materialisms). As a result, the human is relegated to one (albeit very large, very forceful) change-inducing perturbation among many. This kind of thinking is helpful because it denatures our most oppressive social and political orders while inviting scrutiny of how contingency and indeterminacy might produce subjectivity as an "open series of capacities or potencies that emerge hazardously and ambiguously within a multitude of organic and social processes."[5] In short, the "continuous redefining and reassembling of key elements" suggests that the "matter" of any ecology is always in excess of whatever provisionally coherent and stable form it may take.[6] While this excess or vitality is surely active across object–environment relations, the question that remains is how to prevent "the systems that shape us" (e.g., antiblackness, heteronormativity) from creating new "[hierarchies] of living things," now "ranked in terms of their capacity to experience and become anything at all?"[7] I therefore approach the "ecology of the slave ship" as an antiblack signifying formation that organizes our environmental relations, imaginaries, and related desires according to the matrix described above. In doing so, I hope to clarify not only how this "system that shapes us" intervenes in our pursuit of environmental justice, but also how environmental justice might intervene in this signifying formation.

As this agenda indicates, *Toward a Slave Ship Ecology* brings an ecocritical perspective to scholarship on abolitionism and an abolitionist perspective to environmental justice studies. While works like Manisha Sinha's *The Slave's Cause: A History of Abolition*, Andrea J. Ritchie's *Practicing New Worlds: Abolition and Emergent Strategies*, and Jess A. Goldberg's *Abolition Time: Grammars of Law, Poetics of Justice* foreground the movement's radical reorganization of race, gender, and class politics, this book addresses how these politics are developed in relation to the material environment. In this regard, it also builds on the work of environmental justice scholars who trace the correlations between environmental racism, state violence, and socio-political hierarchies. David Pellow's *What is Critical Environmental Justice?* exemplifies this approach by using an interdisciplinary lens to read the Black Lives Matter movement, the US prison-industrial complex, and the Israel-Palestine conflict as issues of and for environmental justice. Similar works include Julie Sze's edited collection *Sustainability: Approaches to Environmental Justice and Social Power* and Harriet Washington's *A Terrible Thing to Waste: Environmental Racism and Its Assault on the American Mind.*

In addition, *Toward a Slave Ship Ecology* is indebted to a body of scholarship situated at the intersections of black studies and the environmental humanities. Prior to this research, accounts of the slave ship were articulated largely from an historical, economic, or sociological perspective, with little attention paid to its environmental or ecological dimensions. The ship is subsumed instead by broader theories about the development of racial capitalism, European colonialism, and the institution of slavery, or treated as one of many (essential) components of the transatlantic slave trade rather than an object or environment in its own right. More recent texts, however, have emphasized how the slave ship's multifaceted constitution exceeds its historical and conceptual boundaries. Christina Sharpe's *In the Wake: On Blackness and Being* employs a tripartite definition of "wake"—as the remainder of a ship's path, a period of morning, and the development of a political consciousness—to demonstrate how the slave ship haunts and illuminates contemporary black life throughout the African diaspora. And Malcom Ferdinand's *Decolonial Ecology: Thinking from the Caribbean World* situates the slave ship as an "imaginary scene" from which to establish an ecological world view based on *marronage*.[8] More broadly, texts like Zakiyyah Iman Jackson's *Becoming Human: Matter and Meaning in an Antiblack World*, Sarah Jane Cervenak's *Black Gathering: Art, Ecology, Ungiven Life*, and Kathryn Yusoff's *Geologic Life: Inhuman Intimacies and the Geophysics of Race* take as axiomatic that natural experience is conditioned by one's position along the color line. *Toward a Slave Ship Ecology* expands on these investigations by concentrating on how the slave ship's multiple registers of meaning influence the development of abolitionism and environmental justice, both historically and intellectually. Subsequently, this book makes clear how powerfully the production of racial difference contributes to our environmental crises. A few additional comments are warranted about several of these texts.

One of the premises of *Toward a Slave Ship Ecology* is that the legacy of the slave ship reverberates through modern configurations of race and nature, a view informed by Sharpe's *In the Wake*. By invoking the term's multivalent meaning, she not only illustrates how and why the abolition of slavery is an unfinished project, but also offers a way of thinking antiblackness and environment together. Her concept of "weather," or a state in which "antiblackness is pervasive *as* climate," refers "not [to] the specifics of any one event or set of events that are endlessly repeatable and repeated, but the totality of the environments in which we struggle; the machines in which we live."[9] In line with this concept, I suggest that miasma theory offers a framework for examining antiblackness as a cumulative environmental effect. Within this framework, antiblackness is figured as the product of multiple race-related stressors that "aggregate and accumulate over time from one or more sources in the built, natural, and social environments."[10]Sharpe's project thus furnishes my own with a concept and method for understanding how antiblackness emerges from our built and natural environments. Kathryn Yusoff's *Geologic Life: Inhuman Intimacies and the Geophysics of Race* extends this understanding to the literal bedrock of modernity, arguing that the earth bears the imprint of racial terror. Yusoff theorizes that race and racial hierarchy emerged geologically, entwined with the colonial history of geology and capitalist resource extraction. Her study recasts geology and geography as imperial projects linked to the hardening and/or reification of racial difference. The surveying, mapping, and mining of lands went hand in hand with the pseudoscientific projects of racial classification. Her reframing of environmental crises as the sedimentation or accumulation of antiblack thought and practice informs my critique of the Anthropocene and its imaginaries.

If Sharpe and Yusoff diagnose how racial slavery and its afterlife structure our material environments, Zakiyyah Iman Jackson presses on the conceptual boundaries of the human that were forged in the same times and spaces. A central tenet of *Toward a Slave Ship Ecology* is that transatlantic slavery established the modern divisions between the human and nonhuman, and between culture and nature. Jackson's *Becoming Human* provides a crucial framework for this argument by showing how Enlightenment thought constructed the human in multivalent opposition to blackness. As she writes, "Black(ened) people are not so much as dehumanized as nonhumans or cast as liminal humans nor are black(ened) people framed as animal-like or machine-like but are cast as sub, supra, and human *simultaneously* and in a manner that puts being in peril because the operations of simultaneously being everything and nothing for an order—human, animal, machine, for instance—constructs black(ened) humanity as the privation and exorbitance of form."[11] *Becoming Human* thus intervenes in contemporary debates about the post-, non-, and inhuman by asserting that we cannot theorize a posthuman future without confronting this "antiblack ontological plasticity."[12] This argument is essential for my analysis of the nonhuman vis-à-vis the slave ship throughout the book.

Each chapter of *Toward a Slave Ship Ecology* corresponds to an aspect of the slave ship's environmental dynamics and more specifically to the relation of its naval architecture to the surplus or deficit of natural resources. Chapters 2 and 3 begin on the high seas and concentrate on how the structural divisions of the slave ship, including the arrangement of the decks and the spaces of the hold, contributed to nascent forms of environmental racism and environmental justice. The ship as a whole functioned as a spatial manifestation of racial dynamics and environmental control. Above deck, crew members moved with relative freedom while raised walkways and open hatches allowed sailors to surveil the hold from above. Below deck, enslaved Africans were crowded onto bare platforms in suffocating numbers. The seams and planks of the hold were sealed with tar and caulking to minimize oxygen exchange, which also facilitated a buildup of pathogens and impeded air circulation. This vertical stratification was key to the ecology of the slave ship. It not only enabled the regulation of water and air to become instruments of power, but also ontologized the boundary between the hold and the deck. This is to say that the black/white binary is informed in part by the slave ship's racialization of space and spatialization of race. Such design choices were neither accidental nor incidental; they emerged from an understanding that within a shared environment, resource management and environmental precarity are effective strategies of race-making.

The management and organization of the slave ship also find echoes in eighteenth- and nineteen-century plantation design and the development of urban cities. Indeed, there is considerable overlap between the ecology of the slave ship and the racial geography and ecology of the plantation. As scholars like Katherine McKittrick, Tiffany Lethabo King, Britt Rusert, and Monique Allewaert have shown, the plantation cultivated techniques of racial, spatial, and environmental management that shaped the American South and influenced our concepts of race and nature. The plantation, McKittrick notes, "uncovers the interlocking workings of modernity and blackness, which culminate in long-standing, uneven racial geographies while also centralizing that the idea of the plantation is migratory."[13] As such, I view *Toward a Slave Ship Ecology* as a supplement to this important work, one that clarifies further how racial slavery and the slave trade established a blueprint for the environmental engineering of race. To draw out these connections, I focus on the two most prominent ships in the history of the transatlantic slave trade—the *Zong* and the *Brooks*, both from the British fleet. Because the atrocities associated with these ships played a decisive role in advancing the abolitionist cause, they occupy a singular position within the environmental imagination as well. More to the point, both ships are notable for the way they draw attention to the ecological features of the slave ship (e.g., resource requirements, sailing conditions, spatial organizations). By investigating how abolitionist representations of the *Zong* and the *Brooks* are taken up and modified within the scholarly literature on environmental justice, I show that resistance to antiblackness has been and continues to be integral to the pursuit of environmental justice.

The fourth chapter explores how the causes and consequences of extreme weather events are displaced onto enslaved black people as the slave ship seeks the shore. Across the Middle Passage, slave ships were perpetually vulnerable to the disorienting effects of gales, storms, cyclones, and other types of inclement weather. The resultant suspension of literal and figurative navigational bearings anticipates, in microcosm, the most unsettling effects of contemporary climate change. In both instances, captive and oppressed communities are made to bear the burdens of devastation and disorientation. I contrast black literary depictions of the destructive force of hurricanes with popular cinematic and journalistic accounts of superstorms and rising sea levels caused by climate change. While black literature links the causes and consequences of hurricanes to the environmental racism borne of slavery's afterlife, I argue that popular cinematic and journalistic accounts problematically employ spectacular images of black suffering to catalyze participation in the movement for environmental justice. The fifth chapter analyzes how attempts to restore balance and stability to the slave ship occurred generally through the production of further environmental deprivations. By withholding or denying access to necessary resources like fresh water, white slavers and sailors constructed a social order that conflated environmental control with racial authority. Accordingly, this chapter juxtaposes the radically democratic, environmentally sustainable response of black speculative fiction to the drought conditions and pollution of twenty-first-century California. In doing so, we can see how efforts to conflate white authority with environmental control betray deep anxieties over the loss of racially exclusive political and economic power. Black feminist authors in particular suggest that these anxieties are rooted in a disavowal of the radical potential of racial blackness to unlock more egalitarian modes of social, political, economic, and environmental organization.

Chapter Summaries

Chapter 2, "The Salt Bones: *Zong!* and the Ecology of Thirst," tracks the development of an "ecology of thirst" across the itinerary of the Middle Passage through a close reading of M. NourbeSe Philip's award-winning 2008 poetry volume, *Zong!* This interconnected series of poems, written about the notorious 1781 massacre aboard the British slave ship *Zong*, suggests that the onboard regulation and distribution of water becomes a crucial strategy for policing the racial boundaries of the human and managing the contested borders between persons, property, and natural resources. Philip reveals that the manipulation of local environmental relations provides a means to institute and naturalize race-making violence by displacing the sources and the consequences of that violence onto the supposed character and constitution of the enslaved. The work of this massive displacement is then mystified so that the effects of antiblack violence—severe impoverishment, fatal illness, forced dehydration—are projected as truths about racial blackness itself, qualities inherent to the black body rather than conditions imposed upon it. It is one of Philip's many achievements in this striking text to demonstrate in form and substance how racialized ecology functioned as a central dynamic of modern slavery from its inception.

Chapter 3, "Bad Air, Built Environments, and the *Brooks*," looks at the infamous 1788 cross-section of the *Brooks*—detailing the brutal arrangement of captive Africans within the hold of this eighteenth-century British slave ship—and at the rich set of writings it has generated in the field of abolition studies in the last generation. This scholarly literature largely discusses the *Brooks* image with respect to its ability to manipulate white racial sentiment in contemporaneous abolitionist debates, but it has so far overlooked the image as evidence of the ecological dimension of the racialization of slavery and its enduring matrix of relations. Using the *Brooks* as a point of departure, this chapter investigates the slave ship as a built environment that anticipates later environmental justice concerns regarding public health, urban planning, and sanitation.

Chapter 4, "Anchoring the Climate Commons: Slave Ship Earth," examines how racial blackness mediates the imagined relationships between the "we" of the climate commons and the "now" of climate change, paying close attention to the strategies used both to represent the climate commons and to solicit investment in its defense. Taking seriously Peter Linebaugh's suggestion to treat the commons as a verb, I first explore the climate justice movement as a form of "commoning" that employs prefigurative politics to create the *communitas* implied by the notion of a commons. I then turn to close readings of two critically-acclaimed climate fiction films, Benh Zeitlin's 2012 *Beasts of the Southern Wild* and Bong Joon-ho's 2013 *Snowpiercer*, arguing that both films anchor their representations of the climate commons on a multiracial community that aspires to and draws from the possibilities of what Stefano Harney and Fred Moten name the "undercommons." Despite their generally inclusive and anti-capitalist ethics, however, *Beasts* and

Snowpiercer represent black suffering as essential to the operations of the undercommons it otherwise champions. In this manner, the films underscore the difficulties of pursuing environmental justice as abolitionism without acknowledging the pervasive, if obscure, libidinal investments in the abjection of racial blackness driving even some of the most radically transformative political projects.

Chapter 5, "The Mattering of Black Lives: Octavia Butler's *Parable* Duology," reads the spaceship as a reconfiguration of the slave ship to determine how Afrofuturism provides new spatiotemporal lenses for understanding slave ship ecology. My reading of Octavia Butler's *Parable of the Sower* and *Parable of the Talents* indicates that the ecology of the slave-ship-as-spaceship initiated a type of "suspended animation" to manipulate environmental experience along the borders of what Hortense Spillers designates as body/flesh. I subsequently claim that the philosophical responses to the Anthropocene, and the new materialisms in particular, are structured by disavowed social fantasies about black female flesh. Butler's *Parable* duology exposes how these responses too often arrogate the radical potential of black women's histories without fully grounding its purview in the socio-political struggles of black people and the critical insights of black radical thought.

To close, the epilogue, "Oceanic Feeling: (Anti)blackness in *The Deep*," examines how the psychic life of race is entangled with the history and materiality of the ocean. To understand fully the ecological impact of slavery and the slave trade requires that we recognize the slave ship as both a "wild and unclaimed richness of possibility" and an "analogy for undifferentiated identity."[14] In locating Freud's oceanic feeling in the Middle Passage, Hortense Spillers provides this mystical sensation with a material history and cultural specificity. I therefore take my cue from Spillers and Rivers Solomon's Afrofuturist novel *The Deep* to sketch a speculative genealogy for oceanic feeling. The novel centers on a community of merfolk, known as the *wajinru*, who are the descendants of pregnant slaves thrown overboard during the crossing of the Atlantic. By reading Solomon's novel alongside the intellectual development of Freud's oceanic feeling, I argue that *The Deep* depicts oceanic feeling as the effect of a therapeutic encounter with black liberatory desire. In doing so, the slave ship and its ecology become a framework for understanding how (anti)blackness animates the power and appeal of the oceanic.

As questions about identity and power continue to shape environmental analysis, representation, and policy, *Toward a Slave Ship Ecology* contributes to these conversations by emphasizing the history of slavery as foundational to our environmental attachments. As such, any reconfiguration of our ethical commitments requires that we acknowledge and address the historical exigencies of the production and reproduction of racial difference. Black literature not only makes clear these demands, but also affords us room to imagine how racial freedom, racial justice, and racial equality might ground a more generous and efficacious environmental ethics. In short, the work of abolition, stretching across centuries, invites us to imagine freedom in ecological terms and against the ecology of the slave ship: where no community is structured by the logic of the deck and the hold and the winds of change and the turning tides carry us toward a more just and livable future.

Notes

1. Manisha Sinha, *The Slave's Cause* (Yale University Press, 2017), 1.
2. Donna Haraway, "Anthropocene, Capitalocene, Plantationocene, Chthulucene: Making Kin," *Environmental Humanities* 6 (2015): 5n.
3. William Lloyd Garrison, "To Our Free Colored Brethren," *The Liberator*, January 1, 1831, 3.
4. M. NourbeSe Philip, *Zong!* (Wesleyan University Press, 2008), 198.
5. Diana Coole and Samantha Frost, "Introducing the New Materialisms," in *New Materialisms: Ontology, Agency, and Politics*, eds. Diana Coole and Samantha Frost (Duke University Press, 2010), 10.
6. Coole and Frost, "Introducing," 14.
7. Donna Jones, *The Racial Discourses of Life Philosophy* (Columbia University Press, 2010), 68; Philip, *Zong!*, 198.
8. Malcom Ferdinand, *Decolonial Ecology: Thinking from the Caribbean World* (Polity, 2022), 21.
9. Christina Sharpe, *In the Wake: On Blackness and Being* (Duke University Press, 2016), 11; 106 (emphasis in the original).
10. "Cumulative Impacts Research," United Stated Environmental Protection Agency, last updated on February 19, 2025, https://www.epa.gov/healthresearch/cumulative-impacts-research.

11. Zakiyyah Iman Jackson, *Becoming Human: Matter and Meaning in an Antiblack World* (New York University Press, 2020), 35.
12. Jackson, *Becoming Human*, 35.
13. Katherine McKittrick, "Plantation Futures," *small axe* 17, no. 3 (2013): 3.
14. Hortense Spillers, "*Mama's Baby, Papa's Maybe: An American Grammar Book*," *Diacritics* 17, no. 2 (1987): 72.

References

Allewaert, Monique. *Ariel's Ecology: Plantations, Personhood, and Colonialism in the American Tropics*. University of Minnesota Press, 2013.

Cervenak, Sarah Jane. *Black Gathering: Art, Ecology, Ungiven Life*. Duke University Press, 2021.

Coole, Diana and Samantha Frost. "Introducing the New Materialisms." In *New Materialisms: Ontology, Agency, and Politics*, edited by Diana Coole and Samantha Frost. Duke University Press, 2010.

Ferdinand, Malcom. *Decolonial Ecology: Thinking from the Caribbean World*. Polity, 2022.

Garrison, William Lloyd. "To Our Free Colored Brethren." *The Liberator*, January 1, 1831, 3.

Goldberg, Jess A. *Abolition Time: Grammars of Law, Poetics of Justice*. University of Minnesota Press, 2024.

Haraway, Donna. "Anthropocene, Capitalocene, Plantationocene, Chthulucene: Making Kin." *Environmental Humanities* 6 (2015): 159–165.

Jackson, Zakiyyah Iman. *Becoming Human: Matter and Meaning in an Antiblack World*. New York University Press, 2020.

Jones, Donna. *The Racial Discourses of Life Philosophy*. Columbia University Press, 2010.

King, Tiffany Lethabo. *The Black Shoals: Offshore Formations of Black and Native Studies*. Duke University Press, 2019.

McKittrick, Katherine. *Demonic Grounds: Black Women and the Cartographies of Struggle*. University of Minnesota Press, 2006.

McKittrick, Katherine. "Plantation Futures." *small axe* 17, no. 3 (2013): 1–15.

Pellow, David Naguib. *What is Critical Environmental Justice?* Polity, 2017.

Philip, M. NourbeSe. *Zong!* Wesleyan University Press, 2008.

Ritchie, Andrea J. *Practicing New Worlds: Abolition and Emergent Strategies*. AK Press, 2023.

Rusert, Britt. *Fugitive Science: Empiricism and Freedom in Early African American Culture*. New York University Press, 2017.

Rusert, Britt. "Plantation Ecologies: The Experimental Plantation in and against James Grainger's 'The Sugar-Cane.'" *Early American Studies* (2015): 341–373.

Sharpe, Christina. *In the Wake: On Blackness and Being*. Duke University Press, 2016.

Sinha, Manisha. *The Slave's Cause: A History of Abolition*. Yale University Press, 2016.

Spillers, Hortense. "Mama's Baby, Papa's Maybe: An American Grammar Book." *Diacritics* 17, no. 2 (1987): 65–81.

Sze, Julie, ed. *Sustainability: Approaches to Environmental Justice and Social Power*. NYU Press, 2018.

United States Environmental Protection Agency. "Cumulative Impacts Research." Last updated on February 19, 2025. https://www.epa.gov/healthresearch/cumulative-impacts-research.

Washington, Harriet A. *A Terrible Thing to Waste: Environmental Racism and Its Assault on the American Mind*. Little, Brown Spark, 2020.

Yusoff, Kathryn. *Geologic Life: Inhuman Intimacies and the Geophysics of Race*. Duke University Press, 2024.

CHAPTER 2

The Salt Bones: *Zong!* and the Ecology of Thirst

a sea of negroes

drowned

live

in the thirst

—M. NourbeSe Philip, *Zong!*

In March of 2014, Detroit's Water and Sewage Department announced that it would begin shutting off water access to clients with outstanding accounts. Given its bankruptcy filing in the previous year, the city clearly saw its massive backlog of about 165,000 delinquent accounts as an opportunity to generate additional revenue.[1] Nevertheless, these water shut-offs disproportionately affected the city's low-income and African American residents. Human rights advocates were quick to point out that these already vulnerable communities were unfairly targeted, considering that water access for commercial and industrial account holders remained untouched, despite the fact that the average debt for industrial clients is more than $10,000 higher than the average debt of residential clients.[2] According to a report filed with the UN's Special Rapporteur on the Human Right to Safe Drinking Water and Sanitation, Detroit's largely poor, largely black populations have historically shouldered the burdens of the city's aging water infrastructure, a situation made worse by economic

D. Leong, *Black Literature, Social Justice, and the Environment*, https://doi.org/10.1007/978-3-032-19288-2_2

depression and a long-standing prioritization of corporate revenue over social welfare. And Detroit is not alone in this prioritization. Various cities in Michigan and the city of Baltimore, Maryland, have also withheld water from black residents, even as surrounding businesses and other residential areas have experienced no interruptions of service. This sharp asymmetry of access is best captured by a single question: What is it like to be thirsty when one is surrounded by water? This question also sits at the heart of M. NourbeSe Philip's 2008 collection of poems, entitled *Zong!*

In 1781, the British slave ship *Zong*, under the command of Captain Luke Collingwood, set sail from the West Coast of Africa toward Jamaica carrying an insured cargo of 470 slaves. Navigational errors en route extended what was normally a voyage of six to eight weeks to over four months, during which time many of the enslaved died from dehydration or prolonged illness. Captain Collingwood, concerned that he would not be compensated for the loss of those who succumbed to these so-called natural causes, instructed his crew to throw overboard and summarily drown to death another 150 slaves.[3] He determined that "the massacre of the African slaves would prove to be more financially advantageous to the owners of the ship and its cargo."[4] When the *Zong* returned to its home berth in Liverpool, the ship's owners filed an insurance claim for the estimated value of the massacred slaves. The insurers denied the claim and the owners subsequently filed suit in a case that was adjudicated in their favor. The insurers appealed in turn and the report of the court's decision to grant a new trial provides the raw material for Philip's collection of poems.

The epigraph from *Zong!* that opens this chapter pits the watery plentitude of the seas against the eternal thirst of the slaves, which compels us to consider how thirst might inaugurate the historical structures underwriting the relationships between blackness and the natural world. This chapter argues that the slave ship is not only a known object in the historic instance, but also an ecology of resource regulations that persists beyond the formal abolition of the transatlantic slave trade. Correspondingly, the poems of *Zong!* indicate that the manipulation of water is part of a long-standing strategy to police the racial boundaries of the human. This transformation of antiblack sentiment into strategies of environmental management affirms that addressing antiblackness is integral to the pursuit of environmental justice. If, as Philip implies, racial slavery initiates and maintains certain relations between organisms and their environments, how has the antiblackness fundamental to that world-historical development informed our ongoing environmental crises? How might the thirst of the slaves also initiate a radical ecology that gestures toward what Sylvia Wynter calls another "mode of being Human"?[5] Philip's exploration of these ecological dimensions is particularly striking when we consider the *Zong*'s role in the abolition of the transatlantic slave trade.

The *Zong* became a powerful symbol within the abolitionist imagination in large part through the efforts of British abolitionists Granville Sharp and Thomas Clarkson. Sharp attended the appeal hearing and commissioned a transcript of its proceedings, which later formed the basis of his own interpretation of the case. While his interpretation was never published as a formal abolitionist tract, it was circulated within abolitionist circles as an indictment of the trade's immorality. As cultural historian Anita Rupprecht argues, the title of Sharp's account—"An Account of the Murder of 132 Negro Slaves on Board the Ship *Zong* or Zurg; with some Remarks on the Argument of an eminent Lawyer, in defence of that inhuman Transaction inclosed in the Letters of 2nd July 1783 to the Lord Commissioners of the Admiralty"—reveals that for him, "the 'transaction' in question was about life, not property."[6] What's more, Sharp's sentimental rhetoric helped to secure the dialectic of the "innocent African" and the "brutal slave trader" that appeared regularly in later abolitionist literature.[7] His response encapsulates what is now considered an essential characteristic of eighteenth- and nineteenth-century abolitionism—a paternalistic condemnation of slavery that presumed whiteness as a universally authoritative perspective.

The conceptual myopia that resulted from this assumption not only shaped abolitionist engagements with the slave ship, but also obscured how racialized ecologies sustained modern racial slavery and continue to sustain its afterlife. It is within this context that Philip's deconstruction of the *Zong*'s legal case opens a spacetime from which to rethink abolitionism and ecology through the slave ship itself. Such a focus not only challenges the "bourgeois sentimentality" of white reformers like Sharp, but also clarifies the connections between the abolitionist pursuit of freedom and environmental justice.[8] *Zong!* thus enables scholars of race and the environmental humanities to examine further how the history of racial slavery is also the history of environmental politics and thought. As Philip's poems indicate, ecology is not only or simply the study of relations between organisms and their environment, it is also a method of meaning-making that foregrounds some relationships while suppressing others. On this score, I argue that the poems of *Zong!* can and should be read as part of the broader archive of abolitionism and environmental justice.

Zong and the Abolitionist Archive

Sharp and Clarkson's publicizing of the *Zong* case granted it a near-mythical status, one in which the *Zong* massacres were established as both a consummate representation of the slave trade and an allegory of moral redemption.[9] Like other mythic figures, the *Zong* derives its power from a certain bracketing of history that "fixes the representation of the past in an ameliorative relation to the national present."[10] As an example, the 2007 bicentennial commemoration of the abolition of the slave trade in Britain featured a replica of the *Zong* ship, which sailed up the Thames River and docked at Tower Pier in London. The Mayor at that time, Ken Livingstone, commented: "It is important that everyone knows the true horror of slavery and the exhibition on board the *Zong* will help educate future generations and give greater understanding of what can only be described as one of the greatest crimes against humanity."[11] The replica also functioned as a floating museum, which featured an exhibit entitled "Free at Last?" that sought to bind contemporary instances of human trafficking—termed "modern day slavery"—to the transatlantic slave trade. The website of the "Spirit of Wilberforce" that sponsored the exhibit through the "Centre for Contemporary Ministry" describes the *Zong* thusly: "The story of the '*Zong*' became a national talking point. The fact that throwing 133 Africans overboard was not regarded as murder but simply as the lawful disposal of 'merchandise' at last stirred the national conscience. Although it took another 24 years to persuade Parliament to ban the slave trade, the '*Zong*' was the first significant turning point in the abolitionist campaign."[12] Both the Wilberforce website and the Mayor's comments imply that even 200 years after the abolition of the trade, the *Zong*'s ability to symbolize the "true horror of slavery" is most valuable as a corrective for national moral degradation. Consequently, the "story of the '*Zong*'" not only confirms the triumph of Britain's "national conscience" over "one of the greatest crimes against humanity," it also serves as a pedagogical tool for "future generations."

While the bicentennial spoke to the enduring influence of the *Zong*, it also gestured toward long-standing anxieties regarding the legibility of slavery and the slave trade. The thorny task of evoking sympathy for slaves without overwhelming white audiences informed abolitionist strategies that protected and reinforced the superiority of white morality. For example, Clarkson's two seminal works, his 1808 two-volume *History of the Rise, Progress, and Accomplishment of the Abolition of the African-Slave Trade by the British Parliament* and his 1786 text *An Essay on the Slavery and Commerce of the Human Species, particularly the African*, dwell on his inability to "find words" that would adequately "exhibit [the slaves'] suffering."[13] As a partial solution, Clarkson constructs an "evocative ethnographic tour" that depicts the slaves' journey from coffle to Middle Passage from the perspective of a fictional "African observer."[14] This narrative strategy not only naturalized (and dehistoricized) the British presence in Africa, but also linked the legibility of black suffering to the moral response of white readers. Responding to claims that his "ethnographic tour" contained descriptions of cruelty too "incredible" to believe, Clarkson implores readers to imagine an "instance" in which "*hundred and thirty two*" slaves were "thrown into the sea."[15] Following this clear allusion to the *Zong*, he asks in a passage worth quoting at length:

> What would the reader have thought on the occasion? Would he have believed the fact...But when he is informed that such a fact as this came before a court of justice in this very country; that it happened within the last five years; that hundreds can come forwards and say, that they heard the melancholy evidence with tears; what bounds is he to place to his belief? The great God, who looks down upon all his creatures with the same impartial eye, seems to have infatuated the parties concerned, that they might bring the horrid circumstance to light, that it might be recorded in the annals of a public court, as an authentic specimen of the treatment which the unfortunate Africans undergo, and at the same time, as an argument to show, that there is no species of cruelty that is recorded to have been exercised upon these wretched people, so enormous that it may not *readily be believed*. If the treatment then, as before described, is confirmed by reason...by what arguments do you *receivers* defend your conduct?[16]

Here, Clarkson frames the *Zong* as both a generalized representation (i.e., "an argument") and a material "instance" (i.e., "an authentic specimen") that nevertheless stands in for the "unspeakable" cruelties of the entire slave trade. This productive tension makes it difficult for readers to dismiss the *Zong* massacres as unusual or exceptional, and catalyzes their conversion from observer to witness. Once readers project themselves into the specific, yet representative scene of the *Zong* massacres, would they be people of "reason" that "bring [these] horrid [circumstances] to light," or mere "receivers" that must deny the "impartial eye" of God to "defend [their] conduct"? This implicit question ties the readers' moral integrity to the suffering of the slaves, so that any reaction other than condemnation appears to let the cruelties of the trade go unchecked and unremarked.

Most significantly, by leveraging the power of empathy to translate the "unspeakability" of the massacres, Clarkson brings "the humanity of the slaves and the violence of the institution…into view via a final allusion to the risk that slaving poses to European bodies."[17] While this transfer of "risk" compels readers to remove any "bounds" to their "belief," it also foregrounds the dangers to white readers at the expense of an actual reckoning with the suffering of slaves. As Saidiya Hartman notes in *Scenes of Subjection*, empathy can

> [confound the witness'] efforts to identify with the enslaved because in making the slave's suffering his own, [the witness] begins to feel for himself rather than for those whom this exercise in imagination presumably is designed to reach. Moreover, by exploiting the vulnerability of the captive body as a vessel for the uses, thoughts, and feelings of others, the humanity extended to the slave inadvertently confirms the expectations and desires definitive of the relations of chattel slavery.[18]

As evidenced by the 2007 bicentennial and Clarkson's published works, the *Zong*'s 132 slaves are most legible against a backdrop of white, virtuous sentiment. The end result is that the deaths of the slaves, not to mention the slaves themselves, are obfuscated by the moral validation and voyeuristic pleasure generated in the performance of empathy. Rather than examining how the Middle Passage re-configured the boundaries of the human for slaves and non-slaves alike, abolitionist representations of the *Zong* trade on the slave's fungibility; as both empty and interchangeable, she is always available as a "vessel for the uses, thoughts, and feelings of others."

If recent and contemporaneous representations of the *Zong* confirm Manisha Sinha's observation that the abolitionist movement is overdetermined by white abolitionists, Philip's intervention is to provide us with alternative approaches to the "unspeakability" of the trade. In an epilogue reflecting on the writing of *Zong!* she asks:

> What did, in fact, happen on the "*Zong*?" Can we, some two hundred years later, ever really know? Should we? These are the questions I confront. Although presented with the "complete" text of the case, the reader does not ever know it, since the complete story does not exist. It never did. All that remains are the legal texts and documents of those who were themselves intimately connected to, and involved in, a system that permitted the murder of the Africans on aboard the "*Zong*."[19]

In rejecting the false equivalency between the "complete text of the case" and the "complete story," Philip points to the unthinkable absence that haunts the archives of the Middle Passage. Because slaves and free blacks were barred from testifying against white persons, what we know about the *Zong* comes primarily from white witnesses, or those "who were themselves intimately connected to, and involved in, a system that permitted the murder of the Africans." The lack of black accounts means not only that the "complete story" of the *Zong* does not and could not exist, but also that "the crimes of slavery must…be confirmed by unquestionable authorities and other white observers."[20] Accordingly, *Zong!* counteracts the tendency of abolitionist texts to substitute white moral authority for black witnessing. By releasing the "cacophony of voices…that had earlier been banned from the [*Zong* case]," Philip instead creates a story which "will tell itself," one that refuses any closure of meaning.[21]

Zong! chronicles the events that occurred aboard that ship in a series of poems gathered into seven interconnected sections. The first of these sections, entitled "Os" (i.e., bone), contains poems constructed only of words found in the text of the *Zong*'s legal case, *Gregson v. Gilbert*. The 26 poems in "Os," like the 26 non-fused bones of the human spinal column, provide the backbone for the rest of the text by highlighting how historical interpretations of the massacres structure our encounters with the *Zong*. "Os" draws our attention to how "much of the language we work with is already preselected and limited, by fashion, by cultural norms – by systems that shape us such as gender and race – by what's acceptable. By order, logic, and rationality."[22] The law, like abolitionism, employs similar "systems" to align the slaves' humanity with "the expectations and desires definitive of the relations of chattel slavery," which produces representations "that [are], at best, only tangentially related to the Africans on board."[23] Even the legal briefs from which Philip works are characterized by the silence of those slaves who remain objects of the case, but not subjects of the law. Hence, work performed in the archive of slavery is "predicated upon impossibility – listening for the unsaid, translating misconstrued words, and refashioning disfigured lives – and intent on achieving an impossible goal: redressing the violence that produced numbers, ciphers, and fragments of discourse, which is as close as we come to a biography of the captive and the enslaved."[24] Instead of straining against this impossibility, Philip pursues a "poetics of fragmentation" that challenges the impulse toward "order, logic, and rationality."[25]

In this regard, the poems of *Zong!*'s remaining sections—Dicta, Sal, Ventus, Ratio, Ferrum, and Ebora—are drawn from a word bank comprised of the legal case and what poet Evie Shockley refers to as "imperfect anagrams" or new words derived from words in the case.[26] What emerges from this bank are sequences of words, phrases, and sounds that gather in clusters or unravel slowly, leaving strings of syllables and letters trailing across the pages. Because "words are broken into and open to make non-sense or no sense at all," these poems are inherently non-linear and even nonnarrative.[27] This entropy increases until the final section, Ebora, features phrases superimposed on other phrases and words running against and into other words. As the reader moves through the poems, she is made keenly aware of her own attempts at meaning-making, many of which are refused by the (anti)structure of the poems themselves. It is only by "un-telling" the dominant accounts of the *Zong*—by disconnecting letters within words, words within sentences, sentences within paragraphs—that the murdered slaves are released from their fixed representations. Philip's deconstructive method allows us to discern what other visions of the human, blackness, and the natural world might exist within and beyond the *Zong*'s legal case and its representation in abolitionist narratives.

"The Order in Destroy"

The *Zong* case played a pivotal role in enabling British commercial law and modern finance capital to ensure/insure the fungibility of the slave. Scholars of its legal proceedings have pointed to the "general average" as the fundamental principle at stake in the appeal. Under the terms of this insurance precept, claims could be made for cargo that was deliberately destroyed if the act was absolutely necessary to secure the success of the larger enterprise. In these instances, all contracted parties (i.e., insurers and the insured) agreed to share in the loss and recompense of jettisoned cargo. In the case of the *Zong*, the ship's owners and crew invoked this clause to make a claim for the average value of the jettisoned slaves by speculating about the value they would have generated in their imagined exchange. But as Ian Baucom points out, the general average is also the "practice by which finance capital insures not only its objects, but, more importantly, its capacity to value (and to *guarantee* the value) of objects regardless of either their thingly existence or their actual market place exchange."[28] The *Zong* case is singular therefore not because it treated slaves as commodities, but because it formalized practices of speculative finance that freed value from the material existence of the object. Value, in other words, no longer depended on the exchange and use value of objects, or even on the actualization of an imagined loss, but was instead conferred in acts of insurance and speculation. "Insurance," Baucom writes, "set the money-form of value free from the life of things."[29] What this means for our understanding of racial slavery is that the edifice of modern finance is built on a violent erasure of black bodies that is not only condoned but *anticipated*, producing "an economy in which blackness circulates precisely as a form of disappearance, a spectral blackness."[30] This circulation of blackness as money-form (i.e., currency) and spectral remainder occurs alongside and through the extraction of labor from the slave's flesh, but is not necessarily bound to it. The slave's flesh is marked as excessive, which, as Philip reminds us, provides the generative locus for a certain legal order.

In "Zong! #2" of the "Os" section, we encounter the following clusters of words[31]:

the throw in circumstance

the weight in want

in

sustenance

for underwriters

the loss

the order in destroy

the that fact

the it was

the were

negroes

the after rains

Wafor Yao Siyolo Bolade Kibibi Kamau

In the context of the *Zong* case, "the loss" belongs to "the underwriters" of the ship's insurance company for whom the systematic murder of slaves represents a contractual obligation. Within this framework, any recognition of the slaves' former bearings, as sons, daughters, friends, and partners, is foreclosed. "Os" reproduces this erasure by rendering the actual personages involved in the massacres—Captain Collingwood, ship owner William Gregson, the sailors, the slaves—as impersonal nouns (i.e., "captain," "negroes," "subject," "underwriters," "mariners," "masters") that recall other abstract legal designations (e.g., defendant, plaintiff, claimant). The individuals of the *Zong* are stripped of historical and cultural specificity as they are forced to occupy legal roles that attenuate the affective and traumatic registers of the massacres. And yet, Philip declares "the Africans on board the '*Zong*' must be named. They will be ghostly footnotes floating below the text … [the] idea at [the] heart of the footnotes in general is acknowledgement—someone else was here before—in *Zong!* footnote equals the foot."[32] All twenty-six poems in "Os" contain a set of African names separated from the formal text by a thin line—footprints of a type of loss that cannot be accounted for within the confines of the law. In her analysis of "Zong! #2," Shockley argues that we should read these names as additional "underwriters" of the text to counter the calculations of value necessary to speculative finance. The word cluster "the weight in want/in sustenance" also draws attention to one homophone that repeats throughout the collection: "want," meaning "to desire or need," and "wont," meaning "one's customary behavior in a situation" or "given to." The word "weight" invokes, first and foremost, the heaviness of those black bodies cast into the sea and the value guaranteed by their deaths. The footnotes at the bottom of the page also bear the weight of lost kinship in want of recognition. In light of the absences that saturate "Zong! #2," and the bar of the law that separates the lost from "the loss," the phrase "weight in want," or the "value in custom," refers to the violences of captivity itself. Within the system of the *Zong* as both text and historical instance, the "wont" (i.e., custom) that provides "weight" (i.e., value) for the "underwriters" of slavery is the eradication of black bodies. Consequently, acts of "destroy" animate a social and legal "order" in which the survival of the slaves is unnecessary for the production of value.

Even though the slaves' treatment as objects of capital was assumed and uncontested, the court overturned the initial awarding of compensation to the *Zong*'s owners and ordered another trial. According to the proceedings of *Gregson vs. Gilbert*, the *Zong*'s slaves were allegedly thrown overboard after the crew concluded that "a sufficient quantity of water did not remain on board…for preserving the lives of the master and mariners…and of the negro slaves."[33] However, during the appeal process, the court discovered that slaves were thrown overboard over the course of several days. During this time, rainfall had supplied the ship with enough drinking water so as to make the further killing of slaves unnecessary. Nonetheless, slaves were still thrown overboard after the rains had fallen, begging the question of what, well beyond the captain's fiduciary self-interest, motivated the additional murders?

The case was clearly never about the ethical significance or legality of these deaths or about the treatment of slaves as commodities. In fact, as Jeremy Krikler proposes, the Chief Justice, Lord Mansfield, may have deliberately "ignored [those] aspects of the law" which contested the legality of the massacres because he was "fixed firmly on the questions of absolute necessity and the general average which he believed might now be weakened if the humanity of the slaves was introduced."[34] Mansfield's well-documented ambition to rid British commercial law of ambiguity certainly guided his decision to safeguard its clarity. Once it became clear that the ship's owners failed to meet the standards of absolute necessity, Mansfield's refusal to consider the slaves' humanity enabled him to shift focus toward the crew and away from the murders themselves. In so doing, the laws of absolute necessity and the general average were left intact, another paramount example of the "order in destroy" found in "Zong! #2."[35] Mansfield's resolve to treat the "unnecessary" murders as nothing more than evidence of crew error works retroactively to confirm the slaves' status as "'empty bearers' of an abstract, theoretical, but entirely real quantum of value."[36] As such, both the law and the systems of speculative finance it scaffolds treat the slave as if she were already dead.

Mansfield's rejection of the slaves' humanity was well founded, given that the laws of slavery relied on narrow definitions of evidence and fact. If the slaves' humanity could "weaken" the law of absolute necessity and the principle of the general average, then its suppression was crucial for the continued operations of the slave trade. "Zong! #9" probes the consequences of this suppression by simulating the processes by which the violences of slavery are rationalized[37]:

slaves

to the order in

destroyed

the circumstance in

fact

the property in

subject

the subject in

creature

the loss in

underwriter

to the fellow in

negro

the sustenance

in want

Rufaro Uwimana Nasiche Nafuna Asura

The "to" placed at the head of the "to the order in" cluster pivots back onto the "slaves" to indicate that the "order in/destroyed" participates in the slave's captive state. The line break between "to the order in" and "destroyed," however, disrupts any pretense that the "order" gained in the destruction of slaves is natural or inevitable. Instead, the reader is obliged to acknowledge the "circumstance in/fact" that motivated such an arrangement. Due to the prohibition against slave testimonials, the experiences of the slaves form no part of the "circumstances" that would bear on the "facts" of the case. In their place, the reader is left with "the property/in subject," so that apprehending slaves as "property" becomes a "fact" that justifies the "order in/destroyed." Without additional facts, then, the reader "[makes] sense of an event that eludes understanding" by turning to culturally familiar "circumstances."[38] And "since we have to work together to complete the events, we all become implicated in, if not contaminated by, this activity."[39] This contamination is vital to the latter half of the poem, or the idea that the "subject" is rendered from the "creature." The species or animal element evoked here makes visible another relation suppressed by the laws of slavery. We can track the creation of the slave from "creature," to "subject," to "property," where each transition requires a more fine-tuned insistence on racial difference. Whereas all "subjects" are made from "creatures," not all "subjects" become "property." Further, because the cluster "to the fellow in/negro" hails the reader directly through the preposition "to," these lines suggest the "negro" is a "fellow" by virtue of a shared condition of being. The reader, like the "negro," is a "creature" too.

"Zong! #9" demonstrates that the law must sever "subjects," "fellows," "slaves," and "negroes" from anything like a shared history or identity so that these "circumstances" can be recalibrated according to the "order in/destroyed." To be sure, abolitionism attempted to undo this severing by stressing the humanity of the enslaved. While this strategy undeniably contributed to the abolition of the trade, it also encouraged a sweeping misrecognition that equated the legal act of emancipation with the desire for, if not the attainment of, racial equality. To observe this misrecognition in action, we need only refer to contemporary claims about America's "post-racial" status, and the systematic roll back of civil rights it has authorized.[40] However, by positioning the law as a metric for racial equality, this misrecognition also conceals the non-juridical mechanisms that preserve and extend the law's promulgation of antiblackness. As such, *Zong!* reveals how this promulgation is continued through the regulation of water.

Ecology of Thirst

Slave ships thrived on an ecology of thirst—a set of relations in which humanity is measured through one's relationship to water. Within these terms, water forms the threshold between slave and non-slave, or between the sailor-as-human and the slave-as-not, as the slave's impossibility of relating to water as "sustenance" bars her from the status of the human. This is to say that the slaves remained thirsty even with an ample supply of water on hand. To be sure, water retains a singular place in any ecological configuration. All living organisms require water to facilitate nutrient transfer (e.g., plants and bacteria), regulate internal temperatures, preserve muscular and cellular structures, and maintain the functions of essential organs (e.g., humans and nonhuman animals). In registering the need for water, thirst is not only an important component of metabolic systems but also (re)orients an organism within their environment. Indeed, it was because of and not despite its commonality across living organisms that thirst became a marker and metric of the "human."

The need for water was a common denominator for all those aboard transatlantic slaving voyages, which obliged sailors to disavow the universality of thirst before it disturbed the social hierarchy of the ship. Because provisions were often in short supply, their distribution was tightly controlled. The practice of placing a "barrel of water and a gun barrel, which was the designated drinking instrument" atop the main sail forced sailors to "climb all the way up to take a single drink."[41] The scarcity of potable water ensured that both sailors and slaves experienced illness or mortality linked to some form of dehydration.[42] Along these lines, the thirst of the slaves became a constant and uneasy reminder of what they shared with captain and crew, a physiological declaration of a common existence. "Zong! #1" uncovers the extent to which this commonality threatened the racial and ideological foundations of the trade[43]:

w w w w a wa

w a w a t

er wa

s

our wa

te r gg g

g go

o oo goo

d

waa wa wa

w w waa

ter o

oh

on o

ne w one

…

w w w w w

a

w wa wa

t

er

wa te

r

wat

er

wa ter

of w

ant

Aba Chimanga Naeema Oba Eshe

The words "want of water," "one day," and "good" are stretched graphically and phonically across the page to form clusters that allow groups of words and letters to relate to each other in new ways. The literal scattering of the word "want" throughout the letters that comprise the word "water" makes it impossible to separate the need from its object because the object itself—water—does not exist independently. Rather, its grammatical coherence is interrupted by patterns of spatial interference that Timothy Morton likens to a vibrating with or attunement to other entities.[44] In these new forms of attunement, we find that the letters of "water" create the space for "want" to mingle with "one," "day," and "goods" to produce new entities. As these entities materialize in the poem's clusters of letters—among them the provocative "ant," "god," "our," and "waste"—they modify the exploded phrase "want of water" in ways that unsettle the restriction of thirst to slaves. "Want of water" becomes more than an expression of lack; it is also a method of translation that reveals how the literal and figurative legibility of water depends on the withdrawal of other entities. "Zong #1!" suggests that the "meaningfulness" of the slave ship's ecological systems depends on "entities that are excluded from the system, yet included by being excluded, thus undermining the system's coherence."[45] The unlocking of "water" and "want" points to what must be excluded from the ecology of the slave ship—a general capacity for thirst—for thirst to constitute the difference between sailor and slave. And as "Zong! #16" makes clear, "should [the sailors] have found/water/&/ being/sufficient," there would have been no need to appeal to thirst to justify the *Zong*'s massacres.[46] If being human requires more than a shared state of "being" and a "sufficient" supply of water, then thirst was a daily reminder of the slave's lack in that regard. As these poems confirm, the legal, social, and political systems leveraged in support of transatlantic slavery were deeply invested in defining the proper mode of being human, one that is emblematized by what Sylvia Wynter calls "Man."

Following the 1992 acquittal of the officers charged with the beating of Rodney King, Wynter penned an open letter to her colleagues that called for a re-examination of how racial blackness has shaped the category of the human. The title of her letter, "No Humans Involved," was drawn from the internal code (i.e., N.H.I.) then used by the Los Angeles Police Department to identify cases involving black males. How, Wynter asks, have we come to "conceive of what it means to be both human and North American in the kinds of terms (i.e., to be White, of Euroamerican culture and descent, middle-class, college-educated and suburban) in which young black males can be perceived, and therefore behaved toward, only as the Lack of the human?"[47] In response, she points to two fallacies around which our conceptions of the human have been structured. The first is that human behavior is driven primarily by the need to secure the survival of the species rather than an imperative to secure the (religious, social, political) conditions necessary for reproducing a culture's "represented conception of the Self."[48] In the second fallacy, this historically and culturally specific representation is mistaken for the totality of the human. As evidenced by the "No Humans Involved" code, this ethno-class version of the human, or "Man," is not only constituted through and against racial blackness, but also establishes the material-semiotic order it needs to "perceive and know itself as if it were a purely natural organism in complete continuity with organic life."[49] Wynter expands on these ideas in a follow-up essay by tracing the overrepresentation of "Man" through the various epistemes of the West, including Greco-Roman and Judeo-Christian philosophies and the natural and human sciences. This category, which we can recognize in its various guises as the transcendental subject, the subject of law, and the subject of civil (and "civilized") society, directs and has directed the form, scale, and scope of socio-political conflicts by measuring the degree to which those involved conform to its requirements. While the category of "Man" has shifted in response to historical pressures, Wynter emphasizes that at least since the advent of modern racial slavery, the mechanisms of "Man" have concentrated on establishing black populations as inferior:

> The correlated hypothesis here is that all our present struggles with respect to race, class, gender, sexual orientation, ethnicity, struggles over the environment, global warming, severe climate change, the sharply unequal distribution of the earth's resources...these are all differing facets of the central ethnoclass Man vs. human struggle...This pattern is linked to the fact that...the Black population group, of all the multiple groups comprising the post-sixties social hierarchy, has once again come to be placed at the bottommost place of that hierarchy...with all incoming new nonwhite/non-Black groups...coming to claim "normal" North American identity by the putting of visible distance between themselves and the Black population group.[50]

This is to say that in order to be or become "Man," one must be non-black. In addition, Wynter's assessment of how these "struggles" are coordinated through "the sharply unequal distribution of the earth's resources" prepares us for understanding how the slave ship's ecological relationships are bound up with the "imposition of ['Man's'] necessity."[51]

In "Zong! #5," Philip restages the legal drama of absolute necessity around the need for water to reveal thirst as the ecological remainder of the slaves' frustrated claims to "Man." Even though rain had more than likely replenished the *Zong*'s water supplies, this was clearly not enough to prevent the crew from throwing slaves overboard. Any additional slaves thrown overboard after the rains reinforced the thirst of the surviving slaves as a *permanent* condition of absolute necessity or need, one that could not be satisfied or alleviated by water. For the slaves, water was a "peril" that could only be misrepresented (i.e., "lying") as "sustenance": "sea and/perils/of water/(one day)/water–/*day one...*/of months/of/weeks/of/days/of/sustenance/lying/ dead."[52] The middle section of "Zong! #5" establishes the permanency of this thirst by making it impossible to ascertain how long the slaves had gone without sufficient water[53]:

of

days

of

sour water

enemies &

want

of

died

(seven out of seventeen)

of

good

(the more of)

of

(eighteen instead of six)

dead

of rains

(eleven days)

of

weeks

(thirty not three)

Odusanya Mxali Ogunba

The quantities scattered throughout this section alternatively measure the dead—"(eighteen instead of six)/dead," the duration of rainfall—"(thirty not three)/of/water," or an unknown element—"(seven out of seventeen)/of." The apparent randomness of the numbers discourages readers from confining the violence of the *Zong* to a specific number—132. Abolitionists likely emphasized the number of slaves thrown overboard because the murder of one slave might prove insufficient to rouse white sympathies. But the focus on this number also reduced the antiblack structures of the *Zong* to a single act of violence so that the abolition of the trade appeared to eliminate any and all conditions through which this antiblackness might be expressed. "Zong! #5" destabilizes this figure by demonstrating the immeasurability, and hence the irresolution, of the captives' thirst. Removing this suffering from the historical time of abolitionist narratives situates it instead within ontological time, or "the time of time itself, the time by which the slave's dramatic clock is set."[54] The slave's non-linear, ontological time enables the historical time of the non-slave to materialize by providing an illusion of stasis.[55] In fact, the slaves' thirst appears to show no historical movement whatsoever[56]:

of

water

day one…

for sustenance

water

day

one…

one day's

water

day

one…

sour

water

day

one...

three butts good

of voyage

(a month's)

Thandiwe Lukman Sabah Liu Sikumbuzo

The consistent italicization of "*day one...*" fixes the fragment as a signifying whole that reaches through the ellipses to span the line breaks. If the slaves' thirst was indeed simply a consequence of the resource scarcity that flourished on slave ships, then a re-distribution of resources should have attenuated the suffering of the enslaved. And yet, as "Zong! #5" demonstrates, even gaining access to water repeatedly returns the slave to "day one" of her journey. If the timeline of the non-slave can be measured by the elimination of thirst, then the slave, defined through her endless thirst, will always remain a slave. This subordination of historical time to ontological time is mirrored in the epigraph from St. Augustine that opens the "Sal" section of *Zong!* : "There was no then."[57] If the slave's thirst does not occupy a specific moment, a "then," it may therefore occupy all moments simultaneously. This racialized experience of water not only became reasonable grounds for treating the slaves as less than human (e.g., the massacres), but also protected the value created by that treatment (e.g., the laws of the general average and absolute necessity). It is hardly surprising that in the middle of its putative "water crisis," Detroit began soliciting proposals for the large-scale privatization of its water and sewer systems. An ecology of thirst guarantees that the ability to control the relationship to water, and moreover to distill value from that relationship, remains one of the hallmarks of "Man." *Zong!*'s second section expands on the effects of infinite thirst to consider how they might propose new ways of being human.

"Sal"

The poems of "Sal" (i.e., salt) position salt as the symbolic demand of the slave, but one with the potential to pose new challenges to the formations of "Man." The ecology of thirst functioned in part by consistently and relentlessly reducing the slave to salt.[58] When the issue of dehydration or thirst arises during a sea voyage, it is almost always attributed to a lack of potable drinking water. But it is also true that salt is a definitive factor in dehydration. Salt allows the body to retain enough fluids to maintain basic bodily functions. When severe sweating or diarrhea occurs in suffocating and toxic conditions, as in the holds of slave ships, bodies lose both water and sodium, making it even more difficult to regulate one's electrolyte balance. If not remedied quickly enough, these losses may lead to hypernatremic dehydration, a condition that is marked by an excess of sodium. This kind of dehydration is not usually the result of increased salt intake (e.g., the ingestion of sea water), and is created instead through consistent water loss that leaves the body unable to manage its usual sodium levels. And it is this excess of salt that creates the need for water. The condition of absolute thirst, however, meant that the slaves' relationship to water surpassed biological need and took on the character of a demand. Or, more precisely, of a drive that was inscribed on the body as a demand, a "corporeal symptom that indicates that the subject's desire has been closed down, and replaced by a somatic formation."[59] What does it mean to experience one's body, and one's mode of being human, as a demand? And to be oriented always toward something that can bring no relief? This convergence of the material (i.e., the body) and the symbolic (i.e., the sociocultural position of the slave) is concentrated in salt, which opens the slave to new forms of interaction that are contained neither by the material nor by the symbolic. Imprints of the slave's existence as salt remain in distributed form, scattered across the seas and beyond the confines of "Man."

"Sal" opens on a mostly empty page that contains two lines: "water parts/the *oba* sobs."[60] In the context of the section's title, "water parts" points simultaneously to salt as the "parts" of (sea) water and the saline content of the oba's "sobs" (i.e., tears), and to the always impending arrival of slave ships. This reference to the Yoruba king, or "*oba*," establishes a connection between the salt of the seas, the violence constitutive of slavery, and the ancestral homeland of some of the enslaved. That this connection is the only one to materialize in the otherwise empty space of the page gives the effect that it has floated to the surface of the undifferentiated mass of nothing surrounding it. Read as either "water parts/the *oba* sobs" or "the *oba* sobs/water parts," there are at least two components to the story "[exaqua-ed]" here.[61] In the first reading, the king's grief is a reaction to salt (salt/the *oba* sobs), and in the second, it is the cause (the *oba* sobs/salt). Taken together, our two interpretations insist on salt's unquestionable imbrication with the horrific events that transpired before, during, and after the slave trade.

As the "bone" of (sea) water, salt is produced through a process of dehydration that also affects slaves in so similar a manner that they find themselves contributing to the saline content of the oceans. The sea became a "cradle" for the perpetually thirsty: "months/for us/of water/for os/in bone/for bone a deep/wa/ter water/deep bo/ne son/g to cradle."[62] "Os" (i.e., "bone") is a close homophone for "us," and when read aloud, these lines can be heard as "months/for os [bone]/of water/for us/in bone." There is both water and salt in the bones of the slave, and the casting of slaves overboard was, in effect, a salting or "boning" of the seas[63]:

on the

ro

se

on

bo ne

on

ne groes

such drab necessity

murder

here we re negroes

like ants

sow the sea

is where

we be seed the seas

with es &

oh & es *os*

In this excerpt, the words "bone" and "negroes" are stacked in such a manner that the syllables of each word are indistinguishable from one another. There is, literally, "bone" in "negroes" and "negroes" in "bone." The slaves "seed" and "sow the seas" with their salt-bones as the water-bone-negroes cluster repeats throughout "Sal," mimicking the triangulation of sea, slavery, and Africa that occurs through salt in the very first stanza of this section. Certainly, the mourning of the oba for salt—"water parts/the *oba* sobs"—is a mourning for those lost *at* sea. At the same time, salt is also the product of a teary or saline remembrance—"the *oba* sobs/water parts"—which ensures that the slaves are not lost *to* the seas.

Curiously, the slaves also seemed to perform the same regulatory function for sailors as salt does for living organisms. Because slave ships were defined by a shared resource scarcity, the sailors had to insure themselves against a too-long, and therefore too slave-like, thirst: "you must hear me/I say/*cum grano/salis*/with a grain of/salt there was in/surance again/st sun."[64] The slaves, like salt, were a constant reminder of the importance of water and in particular of the need to regulate it. For the sailors, the slaves' reduction to salt was a form of "in/surance again/st" the dehydration, or against "the sun," that defined the slaves' material conditions. Considering that the ability to control and regulate water became a defining boundary between sailor and slave, it is tempting to interpret the slaves' existence as salt as the unintended and subversive consequence of an ecology of thirst. Like the return of the repressed, the slaves' symbolic demand forever links the sailor to the slave in a perverse network where the slave-as-salt guards against the sailors' misuse of water supplies. Still, this reading would only be tenable if we underestimate how the ecology of thirst extends, rather than disarticulates, the conditions of captivity: "*le mort le/mort le p tit mort*/scent of mortality/she/falls/*ifàifàifà*/ falling/to/ port/over/ &/over."[65] The death of the slave, her "*mort*," brings a proverbial little death ("*le p tit mort*") or orgasmic release to the sailors, reminding the reader that the sailors' libidinal economy is sutured by the slaves' demand.

As a condition of captivity, the ecology of thirst generates salt as both its cause and its effect. Following the first stanza of "Sal," the reader confronts the repetition of the Yoruba word for "divination"—"ifà"—disseminated across the page: "there is/creed there is/fate there is/oh/oh oracle/there are/oh oh/ashes/over/*ifà/ifà/ifà* i/fa/fa/fa/fall/ing over/&/over the crew/touching there/is fate."[66] While divination refers more commonly to a search for knowledge of the future, it also refers to a search for water. "To divine" or "dowse" is the practice of locating water sources by supernatural insight. The double meaning of "ifà" not only binds the "fate" of the captives to the search for water, but also transforms it into a divine act, a search for futures and fates that confirms the ecology of thirst as an "always an unsettled condition, open to an outside about which it will not know anything and about which it cannot stop thinking, a nervous system always in pursuit of the fugitive movement it cannot afford to lose and cannot live without, if it is to go on existing in and as a mode of capturing."[67] For the ecology of thirst to function as a "mode of capturing," it must maintain a relation between the "inside" of captivity and the "outside" of freedom. Without this relation, the state of captivity would be unrecognizable as such. On this score, the slaves' divine and never-ending search for water supplies the ecology of thirst with its very condition of possibility. By reducing the slave to a "nervous system always in pursuit of [a] fugitive movement," or what I called previously a somatic demand, the ecology of thirst locates the control of natural elements (e.g., salt and water) at the interface between freedom and captivity. In the sections that follows "Sal," we can observe similar processes at work via two other natural elements—wind and iron.

"Ventus" and "Ferrum"

The third section of *Zong!*, entitled "Ventus" (i.e., wind), charts the movement of the *Zong* from the coast of Africa to its arrival in Jamaica, and then to its return to Liverpool. Wind plays a pivotal role not only in each leg of this journey, but also in the visual and linguistic arrangement of the poems. On each page of "Ventus," words, fragments, and phrases are bent into serpentine shapes that recall the movement of the wind across the surface of the ocean. Waves of words and letters cascade down several of the pages while others are blown to the very edges. The arrangement of the poems "suggests something about the relational – every word or word cluster is seeking a space directly above within which to fit itself and in so doing falls into relation with others either above, or below, or laterally."[68] In addition to sculpting the aesthetic form of these fragments, the wind also affects how we understand the relations between fragments by permitting previously disparate elements to intermingle, including those elements shrouded by the overbearing reach of the *Zong*'s legal case.

In the eighteenth and early nineteenth centuries, wind was the principle source of propulsion for ocean-going vessels. The wind that filled the sails of the *Zong* was thus the same that enlivened the slaves' bills of sale: "we seal/the deal/the sale/of/negroes/on/board/the/sail/slap slap in/the wind/some/come from/the fens/others/from the/dales/and/the/far off."[69] The sale/sail homophone prominent in this section acknowledges the relation between technologies that harness the power of natural resources and the slave trade, with the production of value in each instance driven by the slave's fungibility. This relation was apparent even to the sailors: "you take/this negro to be/y our slave we/make good/time the wind/is/with us."[70] The wind also shares with the sailors a point of origin—the dales—which becomes evident as a sailor reminisces about home: "i/come/from/the north/the/dales/land/of mist."[71] This collaboration between slavers and wind is reiterated when the ship's passage up an African river conjures an image of sexual assault: "we/sailed/up/the/cunt/of/Africa/to/found/an/out/caste/race."[72] Linked to the "water parts" verse in "Sal," the *Zong*'s penetration into the continent becomes the founding gesture for the ongoing construction of Africans as the "out/caste/race" *par excellence.* This pivotal moment in the development of modern (anti)blackness is saturated by the violence of those prerogatives (e.g., over Africa and Africans) that the sailors and slavers claim for themselves. The same wind that abets the destruction of the slave's social fabric also delivers the sailors back to theirs: "ru/th cl/air ro/se/ev/e e/va/cla/ra sa/ra/co/ry etc/all/wait/& wait/and/wait/&wait/for/a/ship/to/ bring/their/men/to/them."[73] Listed as the "women who wait" in the ship's fictional manifest, these women are both expectant and expecting: "her/*p tit mort*/turns/from/ruins/of forts/and fortunes/to/found/a/city/on/death/on/ murder."[74] Once reunited with "their/men," the sexual pleasures of "*p tit mort*" transfer the women's attentions from the "ruins/of/forts/and fortunes" in Africa to the preservation of white heritage, or the literal and figurative edifice ("city") founded on "death/on/murder." It is not coincidental that so many of the "women who wait" hail from the origin stories of Western mythology: Circe, Eve, Ruth, and Mary.

Because these mythologies belong to the range of narratives deployed in support of "Man," they are reimagined in "Ventus" as vehicles for the reproduction of antiblackness: "circe argues with eve/about eden/on the eve/of murder…there is no/evidence of eden/in eve eve/to circe."[75] When the Christian tale of creation joins forces with Greek mythology, the link forged from "eve/to circe" offers no "evidence" of peace or "eden." Rather, "the/hag/circe/makes/a/ring/of/stones/in the sand/her o/mens have no/song/or/sound they/sing/of/the/pact/of/pain/be/tween/cain/&abel/bet/ ween/ma/n/&/g/od."[76] In this way, Eve and Circe, Cain and Abel, and "Man" and God are rendered collaborators in the "[founding of]/a/city/on/death/on/murder."[77] To challenge further the supporting narratives of "Man," the penultimate section of *Zong!* sketches an alternative mythology that emerges from within the slave hold itself.

> "Ferrum" (i.e., iron) surrenders to the narrative fragmentation that increases throughout *Zong!* so that the reader must remain at the level of the syllable or word. Yoruba, Spanish, Latin, English, Shona, and French words congregate in dense clusters, with clusters sometimes split between two or three languages. For Philip, the poems in "Ferrum" are unmistakably auditory: "There are times in…*Ferrum*, when I feel as if I am writing code and, oddly enough, for the very first time since writing chose me, I feel that I *do* have a language – this language of grunt and groan, of moan and stutter – this language of pure sound fragmented and broken by history."[78] As the syllables of each fragment drop off and re-connect with other fragments, this rich frenzy of sounds and letters refuses to coalesce into a distinct voice or coherent narrative. In fact, Philip is interested explicitly in these sounds and letters "not working together," or "resisting that order and desire or impulse to meaning."[79] Following her lead, my reading of this section concentrates on how its "language of pure sound" unmakes the ecology of the thirst and the category of "Man."

As an effect of "not working together," the poems of "Ferrum" unmoor the story of the *Zong* from its historical frameworks. The poems' "loss of language and meaning…levels everyone," including the reader, "to a place where there is, at times, no distinction between languages – everyone, European and African alike, has reverted, it appears to a state of pre-literacy."[80] Without the tools of literacy, which enjoin signs to "work together" via the structures of grammar and cultural context, the reader is prompted to build new relationships between signs[81]:

m the deep

w

ale and *sa* *de* e

at fr

esh fish f rom the r

iver we b

e fresh wa

ter neg roes the sea is *ma*

i is *mere* i

s *mer* is

"Ale/aile," for instance, is a homophone meaning "beer" in English and "wing" in French. However, when attached to the "*w*" in the previous line, it becomes the Yoruba word "*wale*," which refers either to a welt on the skin or a wooden plank on a ship. In a similar manner, "*sa*" is a Shona word for "like," until it joins with "*de*" to become "*sade*"—a Yoruba word meaning "honor bestows a crown." In the second half of the poem, the French homophone "*mer/mere*," meaning either "sea" or "mother," precedes the Shona word for "mother"—"*ma/i*."[82] As words and letters tack between different configurations, any attempt at meaning-making is overwhelmed by an abundance of interpretive possibilities. In this respect, the poems draw an analogy between the two "ordering mechanism[s]" at work in the story of the *Zong*: the language and grammar of the law and the category of "Man."[83] These mechanisms together determine the standards by which the massacres are represented and made meaningful and as a result, certain configurations—of signs, voices, objects, natural elements, and bodies – become legible while others are mystified or suppressed. Subsequently, the disabling of these standards draws attention to the manufactured and even "irrational, illogical, and unpredictable" nature of previous configurations while also allowing new configurations to emerge.[84]

What "Ferrum" gestures toward, therefore, is an "undercommon, underground, submarine sociality" made possible by the same "ordering mechanisms" that segregate the slave from the family of "Man."[85] If

"Man" is created by enclosing a small part of the field of possible relations, then these poems demonstrate that a wider range of relations is available to those located outside of the enclosure itself. All along the pages of this section, cats, rats, doves, pigs, apes, corn, hay, rock, iron and other animals, vegetables, and minerals are released through the slaves' "language of pure sound"[86]:

come the tale that can

not be

told in this

tale the *tao*

the way of the dead of

what do

es this mean drat

the rat it

ate the cat or is

it the cat that ate the rat halve

the ration of

cod the globe

Each of these entities are part of the aforementioned "imperfect anagrams" formed from the words of the *Zong*'s legal case. Once freed from their grammatical and categorical confinement—the cat in captain, the rat in ration, the cod in consideration—they become participants in the story of the *Zong* while entering into relations of mutual becoming with the slaves themselves: "ash/es and sa/lt for the bo/die s/of kin un/der the sk/in of s/ea whe/re repo/se the bo/ne sou/ls of kin…part wat/er part bo/ne par/t salt le/sel la sa/l salis in le/sang sa/lt in the e/ye salt i."[87] As an effect of their reduction to "wat/er," "bo/ne," "salt," and "sang" (i.e., blood), the slaves become "kin" to these elements in ways that the sailors are not. These relations, Philip notes, acknowledge that "we are least one and the Other. And the Other. And the Other."[88] To push the terms of this kinship further, the final poem of "Ferrum" recasts the story of the massacres as a tale of resurrection. Doing so ensures that this "submarine sociality" is renewed in every encounter with the *Zong*.

"Ferrum" closes with twenty-two Yoruba, Shona, Akan, Zulu, and Swahili names arranged in an inverted pyramid, some of which reference kinship or the scared and divine (e.g., "Olufunke" meaning "God gave you to me to care for"), while others allude to the week of creation described in the Book of Genesis (e.g., "Kobena" meaning "Born on Tuesday"; "Esi" and "Kwesi" meaning "Born on Sunday").[89] When read alongside the section's epigraph from the Book of Ezekiel, this invocation of the dead is transformed into an incantation for mutual becoming. The verses from which the epigraph is taken recount the "Valley of Dry Bones," or the prophetic vision given to Ezekiel by God in order to inspire the Israelites to return to the holy land. Standing before the titular valley, Ezekiel watches as the bones are rearticulated, covered with layers of sinew and flesh, and reanimated: "There was a noise and behold, a shaking…and the bones came together, bone to his bone…the sinew and flesh came upon them…and the skin covered them above…and the breath came into them…and they lived, and stood upon their feet."[90] The inclusion of this prophecy bookends Philip's earlier efforts to honor the slaves thrown from the ship. The twenty-two names listed in "Ferrum," which are also separated from the formal text of the poems by a thin line, are companions to those cited in "Os." More importantly, the "noise" that initiates their resurrection comes neither from God nor from Philip-as-author, but from those entities unleashed by the poems' "language of pure sound." The slaves of "Os" and "Ferrum" might appear to be barred from direct interaction with the signs, objects, sounds, and entities that populate their pages, but as "part wat/er part bo/ne par/t salt," they continue to (re)animate the ecological configurations of the slave ship and beyond.

Ecologies Undone

While Philip does not claim that the relations deformed and generated across this collection of poems are readily identifiable outside of the text, her focus on how the language of the law and the category of "Man" limit our understandings of the *Zong* reveals that these limitations have material consequences. As evidenced by the water policies of Detroit and other major cities, the antiblack arrangements of the modern era have confirmed that "it [is] now not only the peoples of the Black ex-slave Diaspora, but all the peoples of Black Africa who would be also compelled to confront the inescapable fact... that, as put succinctly by Frantz Fanon, 'wherever he[/she] goes in the world, the Negro remains a Negro' (Fanon 1967)—and, as such, made to reoccupy the signifying place of medieval/Latin-Christian Europe's fallen, degraded, and thereby nonmoving Earth."[91] I hear in Wynter's quotation that the earth has been blackened—forced into an association with all that blackness has come to signify in order to maintain the master code that ranks subjects by their distance from "fallen, degraded, and thereby nonmoving Earth." The farther one is from the natural things of the earth, the farther one is also from blackness. This naturalization of the color line allows "Man" to replace the consequences of regulatory violence with the defects of the less than human, thereby allowing him to act as if antiblackness has nothing to do with the dire ecological circumstances in which we find ourselves.

What I find most compelling about Philip's poetic deconstruction of the *Zong*'s legal case is that its interconnection of water, bone, salt, wind, iron, sailor, and slave (among other configurations) does not stop at an acknowledgment of mutual becomings. The danger of doing so is a certain reliance on the fantasy that interdependence, or the acknowledgment of such interdependence, will be sufficient to prompt new ethical orientations toward human and nonhuman others. Such a hope misrecognizes its own desire to fix race in general and antiblackness in particular as one of many ethical imperatives, capable of being addressed through a shared capacity for self-transformation or a shared humanity. In contrast, the poems of *Zong!* suggest that it is from the position of the slave, or from the black creative and intellectual dispositions it enables, that we are able to grasp the full range of ecological possibilities. Insofar as "Man" delimits our engagements with the natural world, the slave's isolation from "Man" is as much an opening to other relations as it is an enclosure. As Zakiyyah Iman Jackson asserts, "Whether machine, plant, animal, or object, the nonhuman's figuration and mattering is shaped by the gendered racialization of the field of metaphysics" so that "terrestrial movement toward the nonhuman is simultaneously movement toward blackness, whether blackness is embraced or not, as blackness constitutes the very matter at hand."[92] To open oneself to a radical black ecology requires an affirmation of blackness in and as a willingness to be blackened.

Notes

1. Alan Pyke, "Detroit Shuts Off Water to Thousands of Broke Residents," *Think Progress*, Last modified June 20, 2014. https://archive.thinkprogress.org/detroit-shuts-off-water-to-thousands-of-broke-residents-a4056e424137/.
2. A report compiled by The Blue Planet Project for the UN's Special Rapporteur on the Human Right to Safe Drinking Water and Sanitation lists the average debts of Detroit commercial and industrial users as $1,976.98 and $10,817.96, respectively. In contrast, the average debt for individual households is listed as around $540. See "Submission to the Special Rapporteur on the Human Right to Safe Drinking Water and Sanitation Regarding Water Cut-offs in the City of Detroit, Michigan." The Blue Planet Project. Last modified June 20, 2014. https://blueplanetproject.net/2014/06/17/detroit-submission/ "Submission" for more details.
3. While the number 150 comes from the original text of the *Zong* case, *Gregson v. Gilbert*, there is uncertainty about the number of slaves that were onboard the *Zong* and the number that were thrown overboard. Some sources, like James Walvin's *Black Ivory*, list the total amount of slaves murdered at 131, while Marcus Rediker's *The Slave Ship* puts that number at 132. As Philip suggests, however, finding the exact number is most likely impossible because "the ultimate question on board the *Zong*" of "what really happened" is troubled by the unreliable nature of nautical reports and the incomplete archives of slavery. M. NourbeSe Philip, *Zong!* (Wesleyan UP, 2008), 205; James Walvin, *Black Ivory: Slavery in the British Empire* (Harper Collins, 1992); and Marcus Rediker, *The Slave Ship: A Human History* (Penguin Books, 2007), 240.
4. Philip, *Zong!*, 189.
5. Sylvia Wynter, "Unsettling the Coloniality of Being/Power/Truth/Freedom: Towards the Human, After Man, Its Overrepresentation—an Argument," *CR: The New Centennial Review* 3.3 (2003): 263.
6. Anita Rupprecht, "A Very Uncommon Case': Representations of the Zong and the British Campaign to Abolish the Slave Trade," *Journal of Legal History* 28.3 (2007): 334.
7. Rupprecht, "Uncommon Case," 336.

8. Manisha Sinha, *The Slave's Cause* (Yale University Press, 2017), 4.
9. See James Walvin, *The Zong: A Massacre, the Law, and the End of Slavery* (Yale University Press, 2011); Ian Baucom, *Specters of the Atlantic: Finance Capital, Slavery, and the Philosophy of History* (Duke University Press, 2005); Leo Costello, *J.M.W. Turner and the Subject of History* (Ashgate Publishing, 2012); and Fred D'Aguiar, *Feeding the Ghosts* (Ecco Press, 2000).
10. Rupprecht, "Uncommon Case," 344.
11. "Mayor to Open the 'Zong' Replica Slave Ship at Tower Pier," *Greater London Authority*, Accessed March 28, 2007.
12. "The Zong," *The Spirit of Wilberforce*, Centre for Contemporary Ministry, Accessed November 1, 2013.
13. Thomas Clarkson, *The History of the Rise, Progress, and Accomplishment of the Abolition of The African-Slave Trade by the British Parliament*, vol.1 (L. Taylor, 1808), 13–14.
14. Rupprecht, "Uncommon Case," 340.
15. Thomas Clarkson, *An Essay on the Slavery and Commerce of the Human Species: Particularly the African*, (J. Phillips, 1788), 159 (emphasis in the original).
16. Clarkson, *Essay,* 159–160 (emphasis in the original).
17. Rupprecht, "Uncommon Case," 338.
18. Saidiya Hartman, *Scenes of Subjection: Terror, Slavery, and Self-Making in Nineteenth Century America* (Oxford University Press, 1997), 19.
19. Philip, *Zong!*, 196.
20. Hartman, *Scenes*, 22.
21. Philip, *Zong!*, 203.
22. Philip, *Zong!*, 198.
23. Hartman, *Scenes*, 19; Philip, *Zong!*, 199.
24. Saidiya Hartman, "Venus in Two Acts," *Small Axe* 12.2 (2008): 2–3.
25. Philip, *Zong!*, 202.
26. Evie Shockley, "Going Overboard: African American Poetic Innovation and the Middle Passage," *Contemporary Literature* 52, no. 4 (Winter 2011): 808.
27. Philip, *Zong!*, 205.
28. Ian Baucom, "*Signum Rememorativum, Demonstrativum, Prognostikon:* Finance Capital, the Atlantic, and Slavery," in *Victorian Investments: New Perspectives on Finance and Culture*, eds. Nancy

Henry and Cannon Schmitt (Indiana University Press, 2009), 29 (emphasis in the original).

29. Baucom, "*Signum*," 30–31.
30. Jared Sexton, "Captivity, By Turns: A Comment on the Work of Ashley Hunt," *Art Journal* 66, no. 3 (2007): 78.
31. Philip, *Zong!*, 5.
32. Philip, *Zong!*, 200.
33. Philip, *Zong!*, 210.
34. Jeremy Krikler, "The Zong and the Lord Chief Justice," *History Workshop Journal* 64, no.1 (2007): 43.
35. *Philip*, *Zong!*, 5.
36. Baucom, *Specters*, 31.
37. Philip, *Zong!*, 17.
38. Philip, *Zong!*, 198.
39. Philip, *Zong!*, 198.
40. In 2013, the US Supreme Court struck down a key section of the Voting Rights Act of 1965 that required states with pervasive histories of racial discrimination to seek federal approval to modify their election laws. The majority opinion, written by Chief Justice John. G. Roberts Jr., cites the elimination of literacy tests and an increase in African American political representatives as signs that the US has evolved past the need for race-based voting protections (*Shelby County v. Holder*). Shortly thereafter, states like North Carolina and Texas enacted voter registration laws that placed undue burdens on minority residents.
41. Rediker, *Slave Ship*, 206.
42. Kenneth F. Kiple and Brian T. Higgins, "Mortality Caused by Dehydration during the Middle Passage," *Social Science History* 13, no. 4 (Winter 1989): 421–437. See Kiple and Higgins for a more detailed account of the correlation between dehydration and mortality rates aboard slave ships.
43. Philip, *Zong!*, 3–4.
44. Timothy Morton, "An Object-Oriented Defense of Poetry," *New Literary History* 43, no. 2 (2012): 205.
45. Timothy Morton, *Realist Magic* (Open Humanities Press, 2013), 214.
46. Philip, *Zong!*, 28.

47. Sylvia Wynter, "No Humans Involved: An Open Letter to My Colleagues," *Forum NHI Knowledge for the 21st Century* 1, no. 1 (1994): 1–2.
48. Wynter, "No Humans," 4.
49. Wynter, "No Humans," 4 (emphasis in the original).
50. Wynter, "Unsettling," 260–262.
51. Wynter, "Unsettling," 260–262; Fred Moten, "Blackness and Nothingness," *The South Atlantic Quarterly* 112, no. 4 (Fall 2013): 756.
52. Philip, *Zong!*, 9 (emphasis in the original).
53. Philip, *Zong!*, 10.
54. Frank B. Wilderson III, *Red, White & Black: Cinema and the Structure of US Antagonisms* (Duke University Press, 2010), 339.
55. In *Red, White & Black: Cinema and the Structure of US Antagonisms*, Frank Wilderson argues that a fundamental antagonism between the slave and the human generates all the social, historical, and political coordinates of modernity: "If, as an ontological position, that is, as a grammar of suffering, the Slave is not a laborer but an anti-human, a position against which Humanity establishes, maintains, and renews its coherence, its corporeal integrity; the Slave is, to borrow from [Orlando] Patterson, generally dishonored, perpetually open to gratuitous violence, and void of kinship structure, that is, having no relations that need be recognized, a being outside of relationality, then our analysis cannot be approached through the rubric of gains or reversals in struggles with the state and civil society, not unless and until the interlocutor first explains how the Slave is of the world." The slave's "ontological time" is thus also the "epistemological time of modernity itself." Wilderson, *Red, White & Black*, 11, 340.
56. Philip, *Zong!*, 11 (emphasis in the original).
57. Philip, *Zong!*, 58.
58. Meredith Gadsby, *Sucking Salt: Caribbean Women Writers, Migration, and Survival* (University of Missouri Press, 2006); 3. Salt has a long history as metaphor, myth, and material in the corpus of African American and African diaspora literatures. Toni Cade Bambara's novel *The Salt Eaters* is arguably the best-known creative work to consolidate the many dimensions of salt as it depicts Velma Henry's journey to spiritual healing. The title alludes to the myth of the "flying African," who loses her ability to

return to Africa after the ingestion of salt in the Americas. In both the novel and *Zong!*, salt is a poison and an anecdote, a "doubled linguistic sign of adversity *and* survival." Meredith Gadsby's investigations into "sucking salt" as metaphor and physical action insightfully position it as "a strategy for preparing oneself for impending hardship, often in an environment marked by constant upheaval, transition, and economic impossibility. It is a survival skill passed on from generation to generation of Caribbean women."

59. Charles Shepherdson, "The Epoch of the Body: Need and Demand in Kojeve and Lacan," in *Perspectives on Embodiment: Intersections of Nature and Culture*, eds. Gail Weiss and Honi Fern Haber (Routledge, 1999), 203–204. See also Sylvia Wynter, "Towards the Sociogenic Principle: Fanon, Identity, the Puzzle of Conscious Experience, and What It Is Like to be 'Black,'" in *National Identities and Sociopolitical Changes in Latin America*, eds. Mercedes F. Durán-Cogan and Antonio Gómez Moriana (Routledge, 2013). Writing in response to debates over the biological and cultural foundations of psychoanalysis, Charles Shepherdson argues for a return to Lacan's notion of the drive as a somatic demand, or "that material dimension of embodiment that goes beyond organic life, but yet cannot be grasped at the level of speech and intersubjectivity." In his reading, the body is the interface between the biological and the cultural, where drives are always bound to parts of the body because our symbolic networks are inscribed on and through the body itself. Sylvia Wynter's notion of "sociogenetics" reaches similar conclusions through the projects of racialization, noting that antiblack ideologies also operate through the manipulation of bodily experience. Both Shepherdson and Wynter agree this somatic "demand" is not made only at the level of intersubjective speech, insofar as "demand" often emerges in psychoanalysis or critical race studies as primarily a demand for recognition from the Other or the State. As the ecology of thirst suggests, this demand is always for something that, even when given, cannot fundamentally alter the conditions of the slave. It represents instead something like an impasse, or a call for a radical reformulation of our concepts of "nature," "culture," and "race."
60. Philip, *Zong!*, 59 (emphasis in the original).
61. Philip, *Zong!*, 201.
62. Philip, *Zong!*, 68.

63. Philip, *Zong!*, 69 (emphasis in the original).
64. Philip, *Zong!*, 70.
65. Philip, *Zong!*, 62.
66. Philip, *Zong!*, 60.
67. Jared Sexton, "The Social Life of Social Death," *InTensions* 5 (Fall/Winter 2011): 9–10.
68. Philip, *Zong!*, 203.
69. Philip, *Zong!*, 85–86.
70. Philip, *Zong!*, 90.
71. Philip, *Zong!*, 80.
72. Philip, *Zong!*, 97.
73. Philip, *Zong!*, 95.
74. Philip, *Zong!*, 186, 84.
75. Philip, *Zong!*, 84.
76. Philip, *Zong!*, 92.
77. Philip, *Zong!*, 84.
78. Philip, *Zong!*, 205.
79. Philip, *Zong!*, 193.
80. Philip, *Zong!*, 206.
81. Philip, *Zong!*, 169 (emphasis in the original).
82. Philip, *Zong!*, 183–184.
83. Philip, *Zong!*, 193.
84. Philip, *Zong!*, 197.
85. Moten, "Blackness," 742.
86. Philip, *Zong!*, 127.
87. Philip, *Zong!*, 168–169.
88. Philip, *Zong!*, 205.
89. Philip, *Zong!*, 173.
90. Philip, *Zong!*, 126.
91. Wynter, "Unsettling," 319.
92. Zakkiyah Iman Jackson, "Outer Worlds: The Persistence of Race in Movement 'Beyond the Human,'" *GLQ: A Journal of Lesbian and Gay Studies* 21, no. 2–3 (2015): 217.

References

Baucom, Ian. "*Signum Rememorativum, Demonstrativum, Prognostikon*: Finance Capital, the Atlantic, and Slavery." In *Victorian Investments: New Perspectives on Finance and Culture*, edited by Nancy Henry and Cannon Schmit. Indiana University Press, 2009.

Baucom, Ian. *Specters of the Atlantic: Finance Capital, Slavery, and the Philosophy of History*. Duke University Press, 2005.

Clarkson, Thomas. *An Essay on the Slavery and Commerce of the Human Species: Particularly the African*. J. Phillips, 1788.

Clarkson, Thomas. *The History of the Rise, Progress, and Accomplishment of the Abolition of The African-Slave Trade by the British Parliament*, 2vol. L. Taylor, 1808.

Costello, Leo. *J.M.W. Turner and the Subject of History*. Ashgate Publishing, 2012.

D'Aguiar, Fred. *Feeding the Ghosts*. Ecco Press, 2000.

Gadsby, Meredith. *Sucking Salt: Caribbean Women Writers, Migration, and Survival*. University of Missouri Press, 2006.

Hartman, Saidiya. *Scenes of Subjection: Terror, Slavery, and Self-Making in Nineteenth Century America*. Oxford University Press, 1997.

Hartman, Saidiya. "Venus in Two Acts." *Small Axe* 12.2 (2008): 1–14.

Jackson, Zakiyyah Iman. "Outer Worlds: The Persistence of Race in Movement 'Beyond the Human.'" *GLQ: A Journal of Lesbian and Gay Studies* 21, no. 2–3 (2015): 215–218.

Kiple, Kenneth F. and Brian T. Higgins. "Mortality Caused by Dehydration during the Middle Passage." *Social Science History* 13, no. 4 (Winter 1989): 421–437.

Krikler, Jeremy. "The Zong and the Lord Chief Justice." *History Workshop Journal* 64, no.1 (2007): 29–47.

"Mayor to Open the 'Zong' Replica Slave Ship at Tower Pier." Greater London Authority. Accessed March 28, 2007.

Morton, Timothy, "An Object-Oriented Defense of Poetry." *New Literary History* 43, no. 2 (2012): 205–224.

Morton, Timothy. *Realist Magic*. Open Humanities Press, 2013.

Moten, Fred. "Blackness and Nothingness." *The South Atlantic Quarterly* 112, no. 4 (Fall 2013): 737–780.

Philip, M. NourbeSe. *Zong!* Wesleyan University Press, 2008.

Pyke, Alan. "Detroit Shuts Off Water to Thousands of Broke Residents." *Think Progress*. Last modified June 20, 2014. https://archive.thinkprogress.org/detroit-shuts-off-water-to-thousands-of-broke-residents-a4056e424137/

Rediker, Marcus. *The Slave Ship: A Human History*. Penguin Books, 2007.

Rupprecht, Anita. “A Very Uncommon Case’: Representations of the Zong and the British Campaign to Abolish the Slave Trade.” *Journal of Legal History* 28.3 (2007): 329–346.

Sexton, Jared. “Captivity, By Turns: A Comment on the Work of Ashley Hunt.” *Art Journal* 66, no. 3 (2007): 74–79.

Sexton, Jared. “The Social Life of Social Death.” *InTensions* 5 (Fall/Winter 2011): 1–47.

Shepherdson, Charles. “The Epoch of the Body: Need and Demand in Kojeve and Lacan.” In *Perspectives on Embodiment: Intersections of Nature and Culture*, edited by Gail Weiss and Honi Fern Haber. Routledge, 1999.

Shockley, Evie. “Going Overboard: African American Poetic Innovation and the Middle Passage.” *Contemporary Literature* 52, no. 4 (Winter 2011): 791–817.

Sinha, Manisha. *The Slave’s Cause: A History of Abolition*. Yale University Press, 2016.

“Submission to the Special Rapporteur on the Human Right to Safe Drinking Water and Sanitation Regarding Water Cut-offs in the City of Detroit, Michigan.” *The Blue Planet Project*. Last modified June 20, 2014. https://blueplanetproject.net/2014/06/17/detroit-submission/

Walvin, James. *Black Ivory: Slavery in the British Empire*. Harper Collins, 1992.

Walvin, James. *The Zong: A Massacre, the Law, and the End of Slavery*. Yale University Press, 2011.

Wilderson IIII, Frank B. *Red, White & Black: Cinema and the Structure of US Antagonisms*. Duke University Press, 2010.

Wynter, Sylvia. “No Humans Involved: An Open Letter to My Colleagues.” *Forum NHI Knowledge for the 21st Century* 1, no. 1 (1994): 1–17.

Wynter, Sylvia. “Towards the Sociogenic Principle: Fanon, Identity, the Puzzle of Conscious Experience, and What It Is Like to be ‘Black.’” In *National Identities and Sociopolitical Changes in Latin America*, edited by Mercedes F. Durán-Cogan and Antonio Gómez Moriana. Routledge, 2013.

Wynter, Sylvia. “Unsettling the Coloniality of Being/Power/Truth/Freedom: Towards the Human, After Man, Its Overrepresentation—an Argument.” *CR: The New Centennial Review* 3, no. 3 (2003): 257–337.

“The Zong.” *The Spirit of Wilberforce*. Centre for Contemporary Ministry. Accessed November 1, 2013.

CHAPTER 3

Bad Air, Built Environments, and the *Brooks*

> [William Dolben] wished therefore this grievance to be taken into consideration, independently of the general question; and that some regulations, such as restraining the captains from taking above a certain number of slaves on board, according to the size of their vessels, and obliging them to let in fresh air, and provide better accommodation for the slaves during their passage, should be adopted.
>
> —Thomas Clarkson, *The History of the Abolition of the African Slave Trade*

In a 2003 article for the *New York Times*, journalist Helen Epstein set out to uncover why poor urban minority communities have the worst health of any demographic group in the United States. Community members suffer disproportionately from chronic conditions like asthma and diabetes, but no genetic, behavioral (e.g., diet, exercise) or social (e.g., healthcare, income) factor accounts fully for their susceptibility to illness. According to some public health scholars, the more likely cause is a combination of stress and deprivation that seems to emanate from the urban environment itself: "You can't see it, you can't really measure it, but it floats over certain people, especially the poor, and makes them sick."[1] As Epstein notes, this etiology bears a striking resemblance to eighteenth- and nineteenth-century miasma theory which held that diseases are caused by miasmas or "bad air." These noxious vapors were supposedly produced by decomposing organic matter and were thought to accumulate in areas

D. Leong, *Black Literature, Social Justice, and the Environment*,
https://doi.org/10.1007/978-3-032-19288-2_3

of high humidity (e.g., swamps) and population density (e.g., slums). While germ theory would eventually displace these beliefs in the early twentieth century, then as now it proved inadequate for explaining why a correlation between race, disease, and the built environment persists. "Whatever the miasma is that afflicts America's minority poor," she writes, "it is at least partly a legacy of the segregation of America's cities...You'd almost think this new miasma was caused by some sort of infection, because of the way it seems to strike certain neighborhoods and certain types of people."[2] And Epstein is not alone in making these observations.

Over the last few decades, environmental justice activists and scholars have turned to miasma theory to describe the consequences of living and working within racialized environments.[3] In this respect, miasma theory has been employed to link air pollution to respiratory disease, pesticide exposure to cancer, and microplastics to endocrine disruption.[4] For Max Liboiron, this strategy highlights the differences between a particle- and miasma-based model of harm. Much like germ theory, a particle-based model "privilege[s] linear causal links between a discrete pollutant and its pollution" and tends to disregard the many factors that determine when, where, and how a pollutant becomes an agent of harm.[5] This emphasis on linear causality makes it difficult for activists to hold polluters responsible for their damages to environmental or human health. Conversely, miasma theory operates on an "influence model of harm."[6] Because miasmas are dispersed across and inseparable from their environments, their "mechanism of harm [is] not direct, but additive," which means that "everything [is] taken into account and nothing [can] be dismissed or overlooked."[7] Illness is treated in this context as the outcome of several environmental, social, and economic stressors acting in tandem. This model not only shifts disease causation away from individual bodies and toward shared environments but also supports the activist embrace of the precautionary principle. Widely endorsed by scholars and administrators from a variety of fields, the principle calls for precautionary measures when an action, policy, or substance poses a risk to the public or environment, even if a relationship of cause and effect has not been scientifically verified.[8] To this end, a precautionary approach widens the sphere of intervention to acknowledge the possibility of multiple sources of harm. And as Epstein suggests, these sources include the legacies of slavery, segregation, and generational poverty. By rejecting the "pressures to individualize and personalize the problem of why and how particular people suffer from disease," miasma theory attends equally to the effects of illness and its conditions of possibility.[9]

In this regard, miasma theory offers a framework for examining antiblackness as a cumulative environmental effect. Within this framework, antiblackness is figured as the product of multiple race-related stressors that "aggregate and accumulate over time from one or more sources in the built, natural, and social environments."[10] This definition augments existing understandings of racism by foregrounding the complex and often nebulous interactions between the human and nonhuman components of a shared environment. In doing so, it accords with what black feminist scholar Christina Sharpe calls "weather," or a state in which "antiblackness is pervasive *as* climate."[11] The weather, she continues, is "not the specifics of any one event or set of events that are endlessly repeatable and repeated, but the totality of the environments in which we struggle; the machines in which we live."[12] Contrary to this notion of totality, allegations of racism are commonly evaluated on the basis of motivation, intent, or causality. Given that antiblackness is multifactorial, tracing a situation of harm to a discrete cause or "purposeful conduct on the basis of race" is not only challenging, but also can work against the goal of implementing large-scale change.[13] For a stark example, we need only look to the one of the principal rallying cries of the Black Lives Matter movement: "I Can't Breathe."

While the phrase refers explicitly to the final words of Eric Garner and George Floyd, both of whom died by suffocation at the hands of the police, it is also indicative of what Gabriel O. Apata deems a more "systemic and structural suffocation."[14] According to Apata, this "slow and methodical process of social strangulation" occurs through "the withdrawal or denial of vital socio-political and economic means of sustaining life."[15] This description is consistent with much of the response to the murders and the Black Lives Matter movement as a whole. To deflect focus from the institutionalized racism of the police, political commentators and the police themselves reduced the deaths of Garner and Floyd to issues of personal health, behavior, and responsibility. Defense lawyers for Officer Daniel Pantaleo, for instance, cited Garner's "high blood pressure and chronic asthma" to argue that he "caused his own death by resisting arrest," while lawyers for Officer Derek Chauvin attributed Floyd's death to "heart disease, high blood pressure and recent drug use."[16] Similarly, the phrase "I Can't Breathe" is referenced frequently without the activist and scholarly messaging that links it to environmental injustice, racial health disparities, and the organization of urban spaces.[17] By avoiding these concerns, such responses help to ensure that antiblackness, like air, remains "invisible in the manner of its operation and in the finality of its execution."[18]

When we approach the declaration "I Can't Breathe" as a cumulative environmental effect, bad air emerges as an antiblack formation that involves to varying degrees: our ongoing "structural suffocation," what Epstein refers to as "ghetto miasma," eighteenth- and nineteenth-century miasma theory, and the earliest forms of public health and urban planning. In keeping with the precautionary principle, such an approach allows us to address the impact of the formation by identifying the conditions of possibility for, rather than the discrete causes of, its historical and contemporary development. What's more, given the long-standing association between miasma theory, disease, and the built environment, an attunement to bad air can help to clarify how antiblack formations are reflected in the built environment and reified as disparities in health, wealth, and ecological precarity. Indeed, historians often invoke miasma theory to describe the symbolic convergence of public health, urban planning, and the built environment.[19]

The epigraph of this chapter points to an alternative genealogy for this convergence, one for which the slave ship is essential. As I will show, the slave ship concentrated the links between state and commercial interests, health, and social order in a way that not only influenced modern concepts of the built environment but also established the conditions of possibility for bad air to develop as an antiblack formation. The built environment is defined broadly as human-made structures, spaces, and landscapes designed to support human activity. For scholars in the humanities and social sciences, the term also refers to the relationships among these elements and to how these relationships are constructed, managed, and made meaningful. Although the term was not commonly used until the mid- to late twentieth century, it is now applied *ex post facto* to describe the impact of the sanitary movement on urban development during the long nineteenth century.[20] The goal of this chapter, however, is not simply to expand the history of the built environment. Rather, it is interested in how the slave ship's shared environment shaped its cultural and ecological legacies, and in how these legacies are renewed and revised through the built environment's history and conceptualization.[21] The stakes of these issues become clearer when we consider that the most iconic images of the slave trade feature the built environment of a slave ship—the *Brooks*. This chapter therefore explores some of the overlapping social-scientific (e.g., miasma theory, sanitation) and symbolic (e.g., techno-scientific illustration) configurations responsible for the power and popularity of the *Brooks* images. By imagining the slave ship as a spiritual precursor for the built environment and the antiblack formation of bad air, we will understand better how the ecology of the slave ship becomes inscribed in our material spaces.

Following the *Zong* massacre, slave ships retained their prominence within the British national imagination through the circulation of the *Brooks* images. The ship was immortalized in a series of broadsides published by the British Society for Effecting the Abolition of the Slave Trade (hereafter SEAST) and the drawings contained therein were reproduced widely across Europe and the United States. To expose the conditions of transport for slaves across the Middle Passage, each image opens the *Brooks*' hull to detail the arrangement of captive Africans within the hold. Like the *Zong* before it, the *Brooks* made manifest the horrors of slavery for a public with little direct knowledge of the trade. The trend in abolition studies has been to analyze the *Brooks* as a generalized representation of the slave trade, one created for the express purpose of manipulating white racial sentiment.[22] To be sure, the ship was instrumental in galvanizing support for the trade's regulation and abolition. But a full account of the *Brooks*' iconicity must contend with its impact on how we envision the spatialization of race and the racialization of space. I argue that the *Brooks* makes visible a paradox that even now troubles how we perceive the built environment. To be precise, within these images blackness is represented as an emergent property of a spatial arrangement and as a contagion or threat around which the environment must be organized in defense. The *Brooks* thus speaks to racial anxieties over the design and management of urban spaces and shared environments more broadly.

Key to this effect are the styles of illustration to which the images belonged. When we read this visual rhetoric as an expression of *fin de siècle* techno-scientific thought, the *Brooks* appears not only as an archetype of the slave vessel, but also as a mechanical object or machine. Each image that discloses the spatial arrangement of captives in the hold is simultaneously a diagram that schematizes the empire, race, and slave-making functions of the slave ship-as-machine. In this way, the *Brooks* images both naturalize and mechanize the violent processes by which black bodies came to be arrayed within the ship. In cataloguing the dismal conditions of the hold, they also hinted at a parallel between slave ship and city that was grounded in familiar concerns: industrialization (i.e., the slave ship-as-machine), population density (i.e., passengers per tonnage), and sanitation (i.e., the lack of fresh air). Although these concerns were not unique to either space, one factor lent the parallel additional weight. Published in and as abolitionist prints, the images are inseparable from the prospect of a free black population. On the level of the imaginary, the specter of living with or alongside free black communities took the paradox of the *Brooks* off the page and outside of the slave ship.

To plot the alternative genealogies of the built environment and bad air, I first offer a brief overview of the sanitary movement in England and the United States. This section illustrates how the built environment came to embody social and economic stability via miasma theory and practices of hygiene and sanitation. I then track miasma theory through eighteenth-century British maritime and medical culture with a focus on the slave ship as a shared environment. By drawing on medical, pro-slavery, and abolitionist discussions of bad air, I demonstrate how the realities of sharing a racialized environment were displaced onto the logistics of sea travel and the slave ship's built environment. As part of this effort, I review the abolitionist campaign to regulate the slave trade through the Dolben Act. Also known as the Slave Trade Act of 1788, the Dolben Act required slave ships to limit their carrying capacities by tonnage and to employ surgeons to reduce mortality rates among the slaves.[23] It was in response to the Act's passage that SEAST commissioned and published the first drawing of the *Brooks*. After surveying their publication histories, I situate the *Brooks* images in relation to the era's two visual influences: the truth-to-nature model and the mechanical age. These representational modes not only influenced how the images were rendered, meaning in what style of illustration, but also how they were perceived, meaning through which epistemic codes. As I demonstrate, when abolitionist narrative strategies were translated into visual rhetoric, the resulting images displaced practices of race-making onto the physical environment of the ship itself. To close, I return to miasma theory as a way of establishing the slave ship's continuing influence on the built environment.

"All Smell Is Disease"

During his 1846 appearance before the House of Commons Select Committee on Metropolitan Sewage Manure, Edwin Chadwick famously declared that "all smell is disease."[24] Chadwick was already considered an innovator in public health, having published his unprecedented survey of sanitation among the poor four years earlier. As one of the first studies to connect the growth and transmission of disease to living conditions, his *Report on the Sanitary Condition of the Labouring Population of Great Britain* advocated for the improvement of sewer systems, the regular clearing of refuse from streets, and greater ventilation for public and private buildings. It was with respect to these recommendations that he delivered his assertion about smell. Although his work with the Poor Law Commission made him a frequent visitor to Parliament, the occasion for this appearance was a global resurgence of cholera. His statement encouraged officials to prioritize odor elimination, which he concluded would better prevent outbreaks than the purification of drinking water. While we now know that cholera is waterborne, Chadwick's testimony was based on the contemporaneous belief that foul odors were sources and not simply indicators of disease. A Professor H. Booth even declared in an 1844 letter to the architecture magazine *Builder* that it is "from inhaling the odour of beef the butcher's wife obtains her obesity."[25] As his choice of publication indicates, concerns about bad air were increasingly addressed through the design and construction of urban spaces.

During the nineteenth century, European and American cities were growing rapidly, which placed enormous burdens on whatever housing and sanitation infrastructure existed at the time. For the Victorian working classes, overcrowding, starvation, and high rates of poverty and disease were norms rather than exceptions. These dismal conditions motivated many of the era's reforms and raised vital questions about the "condition-of-England" and the nature of public health.[26] Namely, how should the state balance the needs of the working classes with those of commerce and sound governance? By what standards should health be measured? And, to whom does the "public" in public health refer? These questions were as ideological as they were practical and so too were the responses from Chadwick and his fellow reformers. By rejecting "work, wages, and food to focus on water and filth," they sought to align "the moral economy of medicine [with] the political economy of industrial capitalism."[27] Accordingly, the reformers' chief concern was with "*some aspects* of the health of *some* people: working-class men of working age."[28] Miasma theory was pivotal in this regard because it enabled the built environment to absorb the ill effects of industrialization. Instead of addressing how class divisions contribute to poor health and political unrest, reformers fixated on sanitation to combat disease and moral vice (e.g., drinking, gambling). As a result, any illness or behavior could be explained or explained away as the "pathological effect of a bad environment."[29] Without a doubt the sanitary movement's legislative and regulatory measures had long-lasting salutary effects. They not only led to radical improvements in ventilation, sewage, and water delivery systems but also sparked advancements in medicine and the natural sciences. These same measures also reimagined economic and social stability in material terms with an emphasis on infrastructure and environmental conditions. With the help of miasma theory, the built environment came to embody the "condition-of-England" as debates over class and national identity played out on its terrain.

A similar scene unfolded in the United States as policymakers worked to address illness and poverty in the context of industrialization. After taking inspiration from Chadwick, the American sanitary movement developed in response to domestic affairs like the Civil War and Reconstruction. Race was central to this process as scientific understandings of disease and pollution evolved in conjunction with perceptions of race. Like their British counterparts, American reformers used miasma theory to tie morality and hygiene to the built environment. Cleanliness became as a corollary both a physical state and social quality. As cities struggled to accommodate migrants from Europe, Asia, and the rural South, sanitarians supplemented their infrastructure projects with campaigns about proper housekeeping.[30] These programs reflected the assumption that hygiene was primarily economic, or an assortment of behaviors and circumstances related to class. Unlike traits that are given or biological, hygiene could be improved through education and civic engagement. Teaching the public how to maintain sanitary environments thus served the dual purpose of disease prevention and social reform. Still, the problems of urbanization and industrial capitalism were inseparable from anxieties over race and national identity.

In the wake of the Civil War, terms like hygiene, pollution, and contagion acquired new significance. The task of incorporating a now free black population into the nation's socioeconomic and political orders unsettled previous categories of race while also generating new justifications for racism. As the pseudosciences of eugenics and racial pathology gained traction in the late nineteenth and early twentieth centuries, so too did the germ theory of disease.[31] These shifts in scientific thought catalyzed a "turn inward" to bodies, genes, and pathogens, which was represented by the growing belief that "white people were somehow cleaner than non-white people."[32] Now a biological property of race and an expression of class, hygiene helped to reorganize social hierarchies for the postbellum era. Consequently, "contested minorities such as Southern and Eastern Europeans successfully adopted American notions of hygiene to achieve white identity...even as African Americans who engaged in rigorous hygienic practices were denied white privilege."[33] The conflation between race and cleanliness also gave shape to the built environment as reformers drew on fears of disease and contamination to validate exclusionary zoning. The nation's first racial zoning ordinance was enacted in Baltimore in 1910 and was based on the belief that "Blacks should be quarantined in isolated slums in order to reduce the incidents of civil disturbance, to prevent the spread of communicable disease into the nearby White neighborhoods, and to protect property values among the White majority."[34] This is to say that even when black people practiced "acceptable" standards of hygiene, they were treated as contagions or pollutants nonetheless.[35] In this way, racialized notions of human and environmental health were inscribed into the material landscape of American cities.

As this abbreviated history demonstrates, miasma theory was formative to the development of public health and urban planning. Even after the introduction of germ theory, its environmentalist lens continued to inform lay and professional views of health and the organization of cities. In this respect, contemporary miasma theory represents an extension of this trajectory rather than the resurrection of a defunct idea. At the same time, its adoption by environmental justice activists and scholars signals a change in the motivations guiding its use. In the hands of British and American sanitarians, miasma theory played a key role in turning the design and management of the built environment into a mechanism of social control. Doing so helped to ensure that the goals of public health were aligned with the interests of industrial capitalism and an evolving racial order. Conversely, when deployed in service of environmental justice, miasma theory makes this alignment legible to redress its consequences and to disrupt it. These objectives remind us, more importantly, that bad air has long tied the built environment to the production and regulation of social difference. This was especially evident on slave ships.

Race, Disease, and the Slave Ship Environment

Disease prevention was of paramount importance for the eighteenth-century maritime community. While mortality rates on British vessels declined throughout this period, disease remained the leading cause of death for crew members and passengers alike.[36] Outbreaks of typhus, dysentery, malaria, smallpox, tuberculosis, and yellow fever occurred regularly, which compromised the profitability of commercial voyages and the efficacy of naval operations. To mitigate these consequences, maritime surgeons transformed their ships into laboratories for the refinement of medical and sanitary techniques.[37] Their intended beneficiaries were "seamen in the merchant service, crews on naval vessels, and troops travelling to tropical stations" as well as "crews on slave vessels and troops on the West Indies stations," which is to say white British sailors.[38] This distinction is meaningful given that convicts, indentured servants, and especially slaves comprised the majority of passengers on British ships. During the seventeenth and eighteenth centuries, roughly 55,500 convicts and 223,800 indentured servants were transported to the American colonies with another 70,700 indentured servants sent to the Caribbean and 3.1 million captive Africans taken to the Americas, the Caribbean, and Europe.[39] These groups suffered greatly from the noxious conditions of sea travel, but surgeons were interested mainly in protecting the health and labor power of the Empire's sailors. Regardless, the treatments and techniques they developed had far-reaching benefits as they were adopted by military officials, hospital and prison administrators, and later the sanitarians.

Ships hosted specialized disease environments that were ideal for scientific study and experimentation. As they traveled to ports across and around the Atlantic they acquired diverse sets of pathogens that were then carried into periods of isolation at sea. What's more, conditions on board were well suited for the incubation and spread of disease. Heat, dampness, and inadequate ventilation were typical, and the effects of these conditions were compounded by malnutrition and exposure to the elements. This unhealthy and relatively closed environment enabled surgeons to observe the progression of disease, introduce new remedies and procedures, and monitor the results. Naval physician James Lind, for instance, conducted the first clinical trial in modern medicine aboard the *Salisbury* in 1747. His discovery that scurvy is treatable with citrus was the first step toward a major reduction in cases among mariners. Similar achievements include the separation of "fevers" into distinct diseases (e.g., typhus, typhoid fever, malaria), the use of Peruvian Bark to treat malaria, and successful trials of the smallpox vaccine.[40] Although surgeons worked with several theories of disease causation and often in combination, miasma theory had the greatest influence on their investigations. The circumstances that made ships valuable for their purposes also reinforced the idea that bad air is pathogenic.

To clarify how external factors render the body vulnerable to infection, surgeons, and physicians concentrated on the conditions associated with putrefaction or "corrupted" air.[41] Overcrowding, inadequate ventilation, poor sanitation, and improper hygiene all contributed in some way to the onset and transmission of disease. Unsurprisingly, these conditions were found most often in slave forts, workhouses, prisons, hospitals, military encampments, and transport ships. So pronounced were their similarities that typhus was known simultaneously as jail, hospital, camp, and ship fever.[42] This network of sites was not only beneficial for research and testing, but also for illuminating the risks and ramifications of confinement. As medical practitioners traveled to and between sites, they were "testing each other's hypotheses at sea, producing statistics, giving learned papers, writing for medical journals and literary magazines, and publishing books aimed at both a general and a specialist readership," with many hoping to "influence policy both on land and at sea."[43] Like the sanitarians, these reformist doctors were acting mainly to preserve the nation's socioeconomic stability. The author of the first treatise on jail fever, John Pringle, was one such doctor. Published in 1750, his *Observations on the Nature and Cure of Hospital and Jail Fevers* identified the causal matrix for diseases of confinement. Every "receptacle of crowded men, ill aired or kept dirty," he proclaimed, will "produce a fever of a malignant nature."[44] Pringle extended this conviction to military camps and ships in a follow-up publication, noting that the "crowds and the foul confined air of the hold" converted mild fevers into something more "malignant."[45] Both texts emphasized a connection between "pestilence" and jail/ship fever which in his estimation was "worse than the true plague" (i.e., bubonic plague).[46] By reminding readers that a disease of confinement could quickly become an epidemic disease, he built strong support for prison, military, and maritime reform. While this interest in "too many bodies confined in too little space" would go on to define public health and urban planning, it also had a powerful impact on the racialization of disease and the pathologization of race.[47]

For the medical community, miasma theory was part of a broader environmental approach to human difference. As Suman Seth demonstrates, surgeons and physicians relied on a neo-Hippocratic framework to examine variations in physiology and pathology. Within this framework, physical and mental characteristics were considered effects of climate and diet.[48] This interpretation not only shaped medical investigations across the British Empire but also gave rise to naturalistic theories of race. Accordingly, racial difference was regarded as a sign of environmental difference rather than as a manifestation of predetermined traits. To explain why Sub-Saharan Africans were less affected by tropical climates than Europeans, physicians proposed that their skins were better able to "throw out the putrid matter that was so easily produced in warm climes."[49] While naturalistic theories tended to portray dark skin as advantageous, they also provided authoritative rationales for slavery. Plantation owner and historian Edward Long wrote in 1774 that slaves in Jamaica "labour with most alacrity and ease to themselves in the very hottest part of the day" because the "openness of their pores gives a free transpiration to bad humors."[50] His claim reflected a belief popular among physicians that Africans were "less inclined to bilious diseases" because their constitutions were "'more robust' than that of whites."[51] This "putrefactive paradigm" was central to most accounts of racial difference until abolitionists began to challenge its logic.[52] If Africans were "better fitted by nature" for slavery, they asked, why were slaves suffering from such high rates of morbidity and mortality?[53] And why were Europeans laboring in the same environments often healthier and "more robust" than slaves? Indeed, these contradictions were nowhere more clear than on slave ships.

The profitability of a slaving voyage was tied directly to the health of its captives and for this reason surgeons had been employed on slave ships since the beginning of the century. By pursuing together the maximization of profit, the preservation of health, and the maintenance of racial hierarchy, they became powerful instruments of racial capitalism. Like captains and vessel owners, surgeons were compensated according to the number of slaves brought successfully to market.[54] Thomas Aubrey advised his fellow surgeons in 1729 that when practicing on a slave ship, "the more you preserve of them for the Plantations, the more Profit you will have, and also the greater Reputation and Wages another voyage."[55] Given this pay structure, surgeons were incentivized to implement preventive measures for disease. These included hygienic programs for the crew and slaves, dietary and exercise regimens, and improvements to sanitary facilities.[56] Historians of the trade have pointed out that these measures were successful largely because surgeons' "sophisticated understanding of the transmission of disease [came from] reformers in the prison, convict, and hospital services."[57] Slaving vessels carried more bodies per ton than any other kind of ship, including those transporting laborers, troops, prisoners, and migrants.[58] As the epitome of "too many bodies confined in too little space," they were able to benefit from the treatments and techniques developed for comparable situations. Despite these similarities, many surgeons were reluctant to connect these conditions of confinement to the suffering of the slaves.

As debates over slavery intensified in the mid- to late eighteenth century, the slave ship became a formative site for racial essentialism.[59] Medical practitioners played a substantial role in these debates as abolitionists and pro-slavery advocates called on their expertise or sought to dispute it.[60] The field of medicine, in turn, recalibrated its approach to racial difference as abolitionists drew attention to the cruelties of the slave trade. Chrisopher Blakley has shown that after the Seven Years' War, doctors increasingly "held that confinement, filth, and unventilated air posed medical risks to White sufferers but did not particularly endanger Black people."[61] Physician Charles Bisset, for example, condemned the overcrowding, heat, and putrid "effluvia" of jails and ships while proposing that Africans "whose solids are very compact, rigid and elastic, are little affected by those humid vapours in the warmest air."[62] James Lind wrote extensively about the dangers to Europeans of the "confined and corrupted Air in a large crowded Ship," even as he argued that slavery was an "absolute necessity" because Africans were better suited for warm climates.[63] The Deputy Inspector of Hospitals and Fleets, James Veitch, acknowledged these contradictions openly by admitting "if infection…is produced only in situations wherein many are crowded together in a comparatively small space," then "it becomes a question of no small importance, why malignant and infectious fevers are never, or very seldom, generated on board Slave Ships." Notably, he provided no evidence for this near absence of fevers before crediting surgeons with their prevention and then insisting, rather paradoxically, that African constitutions were neither "predisposed to" nor the "subjects of acute fever."[64] As these examples indicate, medical practitioners sought to "[naturalize] African confinement and incarceration aboard ships in the Atlantic world as comparatively salubrious and normal, while raising the alarm of the threat posed by crowding and confined spaces for

Europeans."[65] In doing so, they were able to claim that the slaves' poor health was caused not by their conditions of captivity but by some innate aspect of their biology. The idea that black and white bodies have fixed "sensibilities" further suggested that slaves required different standards of care.[66] What appeared to Europeans as substandard living conditions were actually proper to African physiology. In this respect, the slave ship accommodated the transition between naturalistic theories of race and the hardened concepts that prevailed in the nineteenth century.

During this period, abolitionists continued to denounce the slave trade via essays, pamphlets, public testimonies, and eventually the *Brooks* broadsides. Against the medical trend of naturalizing confinement, they characterized the slave ship's environment as a precipitating factor for disease. In his influential *An Account of the Slave Trade On the Coast of Africa*, surgeon Alexander Falconbridge concluded that the primary cause of illness among "much crowded slaves" was "the breathing of a putrid atmosphere." As long as that cause remained, he asserted, even a surgeon's "greatest skill and attention" could do little to alleviate their suffering.[67] Unlike many of his colleagues, Falconbridge placed the blame squarely on overcrowding and neglect for the slaves' poor health. By emphasizing the inability of surgeons to address these causes of illness, he maintained that the problem lay not with the slaves' biology but with the design and management of the ship itself. Captain John Newton corroborated these allegations, writing that due to a "want of room" and lack of cleanliness slaves were forced to "breathe a hot and corrupted air." Consequently, "epidemical fevers and fluxes which fill the ship with noisome and noxious effluvia often break out" and "infect the Seamen likewise."[68] Here, Newton verified that confinement posed a risk to all on board by highlighting the sailors' vulnerability to disease. Surgeon Thomas Trotter, who served on the *Brooks* in 1783, testified to a Parliamentary Committee that he "observed [the slaves] draw their breath with all those laborious and anxious efforts of life, which we observe in expiring animals." When asked if the "slave rooms" were sufficiently ventilated, he responded, "When the scuttles are obliged to be shut, I do not think the gratings are by any means sufficient for airing the rooms; for I never could myself breathe with freedom, unless I was immediately below the hatchway."[69] As these reports illustrate, the slaves' conditions of confinement affected most aspects of a voyage. From impacting the health and work of the crew to determining the allocation of space, they confirmed beyond a doubt that the slave ship was a shared environment. This is to say that while the ship itself was compartmentalized, the social and material repercussions of captivity were not. In this light, Trotter's reference to "breathing with freedom" takes on new resonance.

The questions of how and what it means to inhabit a shared environment were foundational to the slave ship and to abolitionist debates. While mostly implicit or implied, these questions were as much about the production of racial difference as they were about resource management and moral responsibility. I argue in the previous chapter that the regulation of water became a strategy for policing the racial boundaries of the human. Because water is a common denominator of survival, sailors were obliged to disavow the universality of thirst before it disturbed the hierarchy of the ship. In this context, the ability to "breathe with freedom" is analogous to thirst, with the need for air not only uniting sailors and slaves but also informing attempts at its regulation.[70] Unlike water, however, air could not be stored, distributed, or constrained, or at least not to the same extent and with the same facility. By its very nature air is unruly, expansive, and most importantly, shared involuntarily. Although sailors slept, worked, and lived on deck and separately from the slaves, interactions between the two groups were frequent. Slaves were brought on deck for a variety of reasons including exercise, bathing, and "entertaining" the crew, and crew members descended into the hold to clean and enforce "discipline."[71] Moreover, while the "slave rooms" were physically isolated from the rest of the ship, a collection of portholes (i.e., scuttles), gratings, hatchways, and wind-sails allowed air to circulate between the hold and the deck.[72] Even after the introduction of hygienic programs, this combination of frequent interaction and air exchange often facilitated the spread of disease instead of preventing it.[73] Like Newton's ship full of "noisome and noxious effluvia," it seemed as though bad air was everywhere. This impression certainly was not limited to the slave ship, but its entanglement with race made it troublesome in distinct ways. Naturalized or not, the slaves' conditions of confinement generated a miasmatic atmosphere that transgressed the boundaries between the hold and deck, and sailors and slaves. The repercussions of this transgression were not only morbidly apparent, they also amplified a reality crucial to the antiblack formation of bad air. By this I mean that in some way and to some degree, an environment is always shared involuntarily.

The pro-slavery, medical, and abolitionist discussions of bad air may have taken different approaches to the slaves' confinement, but they ultimately served to regulate race and space within shared environments.

The ability to define bad air, determine its conditions of possibility, and recommend a course of action was related intimately to the management and organization of racial difference, both on the slave ship and beyond. This regulatory relationship emerged most clearly during Parliamentary debates over the Dolben Act. After the *Zong* massacre in 1781, the abolitionist movement gained significant momentum, which led to the founding of SEAST in 1787. In May of the following year, William Dolben called on Parliament to take immediate steps to regulate the slave trade. To introduce his legislation, he expressed equal concern for slaves and sailors, both of whom "perished" frequently as a result of the holds' "contagion."[74] If no "remedy" was applied, he warned, "no less than ten thousand lives might be lost between this and the next session."[75] As evidenced by his focus on the "number of slaves," "putrid and fatal disorders," and "fresh air," the Act reflected the abolitionist arguments that overcrowding, medical neglect, and inadequate ventilation turned the slave ship into a place of "solid pestilence."[76] To ascertain if the proposed regulations would reduce the trade's high mortality rates, Parliament solicited testimony from experts and witnesses familiar with the slaves' conditions of confinement. Proponents of the trade argued that regulation was unnecessary because the slaves were healthy and happy and their ships were well ventilated. The Baron of Heathfield, General George August Eliott, submitted that "the amount of air breathed by the Africans on Middle Passage compared favorably with the amount consumed by soldiers in their tents."[77] Captain Robert Norris spoke of the slave ship's "spacious gratings," and the "Row of Air Ports all round the Sides of the Ship, to admit a free Circulation of fresh Air."[78] To this he added that overcrowding was not a problem because "warmth is so acceptable to negroes, that when a ship had only half its complement…on board, those negroes then there lay as close to each other by choice."[79] In keeping with the naturalization of confinement, these testimonies implied that the hold's bad air was in fact not bad air at all. By citing the slaves' preference for "warmth" and the supposed absence of fevers to which they were vulnerable, pro-slavery advocates implicitly identified the atmosphere of the hold with the climate of Africa.

As abolitionists worked toward the prohibition of the trade, they advanced a similar argument albeit on behalf of sailors rather than slaves. The Slave Trade Act was passed into law in July 1788 but because it expired after a year, Parliament was required to vote on its extension annually. Subsequently, during deliberations over the Act's renewal, abolitionists repeatedly stressed the limitations of regulation. They concentrated their efforts on two issues in line with Dolben's initial framing: the slaves' mode of transportation and the sailors' health and well-being. As to the first issue, they argued that because the slaves' transportation "necessarily exposed them to many and grievous sufferings," no regulation of space, medical care, or air quality could transform the hold into something healthy and humane.[80] To illustrate this point, they built a campaign around images of a slave ship's interior—the *Brooks*. I examine this campaign at length in the sections below, and to discern better its relation to race and space, it is helpful to see how that relation unfolded in the interests of white sailors. Plainly, sailors were not the subjects of abolitionism, but they were sympathetic in ways that the slaves were not. To counter the assertion that the trade was a "nursery for British seamen," abolitionists quickly foregrounded the African environment as harmful to sailors.[81] Just as the slave ship environment was a precipitating factor for disease, the African environment likewise became a place of "solid pestilence." This perspective was beneficial for promoting abolition over regulation because anything less than the dissolution of the trade would expose sailors to the "unhealthiness" of Africa's "inhospitable climate."[82] As it stood, abolitionists claimed, sailors were "caught between the toxic interior environment of the slave hold and the 'excessively hot' and humid environment of the West African coast."[83]

The movement's most prolific writer, Thomas Clarkson, elaborated on this predicament in a text entitled *An Essay on the Comparative Efficiency of Regulation or Abolition, as Applied to the Slave Trade*. The essay methodically refutes pro-slavery arguments while inviting readers to weigh the benefits of regulation versus abolition relative to sailors and slaves. The section concerning sailors identifies "boatwork, insurrection, and disease contracted from the disease of the slaves" as their chief causes of death, all of which were "peculiar to the slave trade."[84] According to Clarkson, boatwork, or the practice of taking small boats upriver to procure captives, was manifestly "prejudicial to the lives of seamen" because it exposed them to the "inclemency of the weather night and day."[85] This exposure continued unabated even with the slave ship at harbor, as sailors were "never permitted to sleep between decks" yet were subject to the "noxious steam and effluvia" emanating from the hold.[86] Clarkson concluded that for the sailor "who woods and waters in Africa, who works up to his knees, and frequently up to his breath in swamps, who is out in an open boat for a fortnight at a time, and who is subjected on board to the stench and disorders of a number of miserable beings," the trade was not a nursery but a "grave."[87] Even though it would take another eighteen years for the trade to be abolished, this line of thinking made an immediate impression on members of Parliament. The essay was published in 1789, and Clarkson was able to inject its arguments and evidence into discussions about the renewal of the Dolben Act.[88] The revised Act took seriously his criticisms of the sailors' lack of shelter and their prolonged contact with the African climate. To that end, its new provisions required captains to hire "natives" to "wood and water" their ships, to provide an "awning" for sailors involved in boatwork, and to reserve space for at least "half the Ship's Company to shelter themselves while the Slaves are on board."[89] By carving out a space for sailors in the African and slave ship environments, these provisions symbolically (re)compartmentalized each environment along racial lines.

The Dolben Act constituted a formal acknowledgment that the slave ship was defined by its built environment. While this fact was hardly surprising to abolitionists or pro-slavery advocates, it also gave form and shape to uncertainties over race. As a result, the complexities of sharing a racialized environment were negotiated through the logistics of sea travel (e.g., disease prevention) and the design and management of the ship's built environment. We can recognize this displacement at work in the pro-slavery and abolitionist contributions to the Act's debates. In each case, the discourse of bad air mapped the symbolic boundaries of race onto or through the physical components of the ship. For pro-slavery advocates, bringing the climate of Africa into the hold not only racialized captivity itself, it also fortified the imaginary border between the hold and the deck. Abolitionists, by contrast, extended the atmosphere of the hold to the climate of Africa. The bad air of the slave ship was no longer only or primarily a product of confinement; it was now also an inherent characteristic of a racialized environment. As Paul Sampson points out, because "the locus of aerial danger had shifted from the interiors of crowded ships to the hot, insalubrious climates that formed a growing part of Britain's empire," the new goal "became to insulate British bodies from inhospitable environments – a rationale that gradually evolved into the insulation of certain people from others."[90] Accordingly, the Act's new provisions reorganized the interior of the slave ship and its river boats to protect white sailors from the perils of these environs. While the push for insulation is significant on its own, what makes it more remarkable is the fact that it was accompanied by the *Brooks* broadsides and their images of slaves in isolation. As I suggest below, the idea that the built environment can simultaneously produce and contain racial difference was key to the abolitionist project. Regardless of intent or motivation, the *Brooks* images, discourse of bad air, and shared environment of the slave ship formed the conditions of possibility for what would eventually become the antiblack formation of bad air.

"Some One Ship"

The *Brooks* was built in Liverpool in 1781 for slave trader Joseph Brooks Jr. and at 297 tons was considered large by eighteenth-century standards. During its commission, it completed ten voyages and carried an estimated 5163 captives, 4559 of whom reached their destinations alive.[91] SEAST selected the *Brooks* from a list of slave ships measured and documented by Captain Parrey as part of the effort to pass the Dolben Act. During Parliamentary debates about the bill, William Pitt sent Parrey to Liverpool to obtain more information about transport conditions across the Middle Passage. Following the Act's adoption, SEAST was determined to show that while a welcome development, the new regulations did little to alleviate the suffering of slaves. "It must be obvious," Clarkson wrote, "that it became the committee to select some one ship, which had been engaged in the Slave Trade, with her real dimensions, if they meant to make a fair representation of the manner of the transportation."[92] The *Brooks*, which Parrey concluded was "grossly overcrowded," was selected in part because the use of its real dimensions would reveal "regulation itself as perfect barbarism."[93] The overall effect of this first image was and still is striking.

The first image of the *Brooks* was a copperplate engraving designed by artist and scientist William Elford on behalf of the SEAST Plymouth Committee in 1788. Entitled "Plan of an African Ship's Lower Deck with Negroes in the proportion of only One to a Ton" (hereafter the Plate), it features one aerial view of 294 slaves placed in four separate compartments: "Girls Room," "Women's Room," "Boys Room," and "Men's Room" (Fig. 3.1). With only six feet by sixteen inches allotted for men, five feet by fourteen inches for boys, five feet, ten inches by sixteen inches for women, and four feet by fourteen inches for girls, the slaves are presented lying in orderly rows that run the length of the ship. Drawn entirely in black and set against a white background representing the ship's cargo hold, the slaves appear simultaneously as a faceless mass and individual figures with unmistakably human bodies. Heads and legs are readily identifiable even as the slaves' torsos appear joined in uninterrupted lines. Like "herrings in a barrel," they are arranged so that every inch of available space is utilized, an achievement underscored by the crude symmetry of the image itself.[94] Bleak and overwhelming, the Plate gave visual form and visceral force to the phrase "living cargo."

The public appearance of the Plate marked an innovation in eighteenth-century print and visual culture. While abolitionist tracts, books, and essays were already in circulation, they generally avoided illustrations beyond the occasional map or chart. When the Plymouth Committee attached the Plate to a four-page pamphlet, it created the first integration of text and image in an abolitionist publication (hereafter the *Plan*). As Cheryl Finley observes, the inclusion of an "explanatory illustration (or having long explanatory remarks for an illustration) would soon revolutionize the way that abolitionists conducted their campaign of garnering public support."[95] The Plymouth pamphlet included two images, the Plate and an imprint of Josiah Wedgwood's abolitionist medallion featuring a chained and kneeling slave who asks, "Am I not a Man and a Brother?" These were accompanied by eight paragraphs written by Elford that invite the reader into the imaginary and brutal world of the Plate. The first four paragraphs detail the space allocated to each slave, their fetters, feeding routines, and overall transport conditions, while the final four refute popular pro-slavery arguments, including the arguments that the trade served as a "nursery" for British sailors and that its regulation would lead inevitably to the abolition of slavery. The

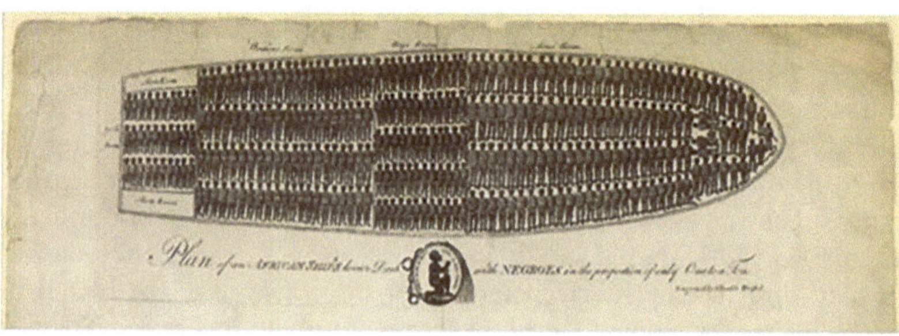

Fig. 3.1 Plan of an African ship

sixth paragraph, for instance, assures readers that "liberating the Slaves forms no part of the present system; and so far will the prohibition of a future trade be from injuring private property, that the value of every Slave will be very considerably increased."[96] Both the Plate and *Plan* were extensively distributed and well received, so much so that within a few months the Philadelphia-based New Society for Promoting the Abolition of Slavery published their own versions in the magazine *American Museum*.

While the Pennsylvania and Plymouth versions differed in size and form, their most substantial differences lay with their text. Pennsylvania's first version debuted in the May 1789 issue of *American Museum* as a two-page magazine article with the Plate attached as an insert. The Society swiftly authorized a second printing run with a request for copies that were easier to distribute. As a result, text and image were consolidated further so that the second version could be published as a broadside (Fig. 3.2). In addition to this physical evolution, both versions carried the same revisions to Elford's original text which centered on an explicit anti-slavery stance. Titled "Remarks on the Slave Trade," the broadside removed Wedgwood's medallion along with the sixth paragraph disavowing support for emancipation. This paragraph was replaced with a new introductory paragraph that credits the "Pennsylvania society for promoting the abolition of slavery," before transitioning to an interpretation of the *Brooks* image: "Here is presented to our view, one of the most horrid spectacles – a number of human creatures, packed, side by side, almost like herrings in a barrel, and reduced nearly to the state of being buried alive, with just air enough to preserve a degree of life sufficient to make them sensible of all the horrors of their situation."[97] Although the paragraph was never reproduced outside of the Pennsylvania broadside, the broadside itself would become the medium of choice for abolitionist materials going forward.

Through its next round of revisions, the *Plan* evolved into the set of images that were and are most prominently associated with the slave trade. Published by the London SEAST committee as a broadside in April 1789, the "Description of a Slave Ship" (hereafter the *Description*) revised Elford's original text and offered six additional illustrations of the slaves' quarters (Fig. 3.3). These included a bow-to-stern longitudinal cross-section indicating the placement of at least six decks, two transverse cross-sections of the vertical arrangement of decks and slave-holding platforms, and three more aerial cross-sections, two of the holding areas near

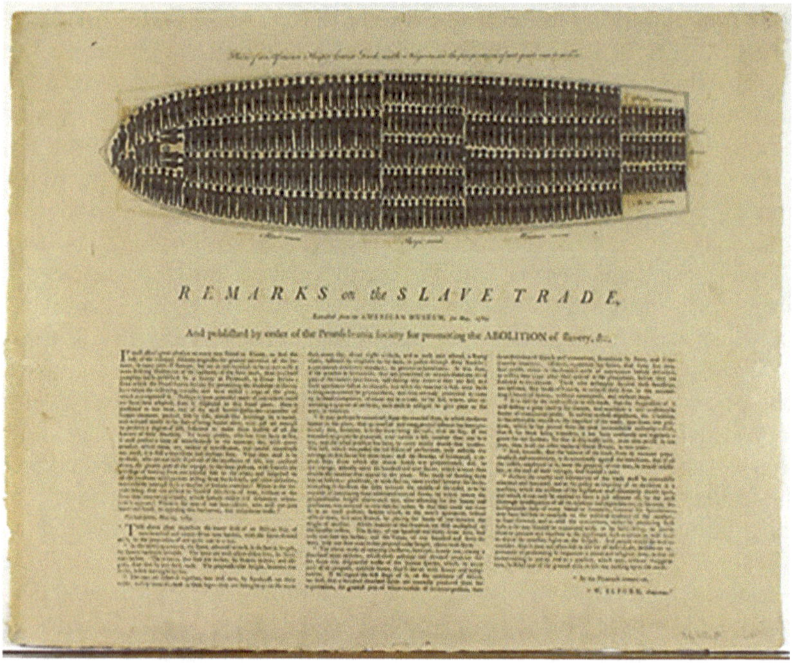

Fig. 3.2 Remarks on the Slave Trade

the stern of the ship, and one which displayed how slaves were placed on a platform running above the cargo deck of the original illustration. Each illustration is identified by a figure number (e.g., "Fig. I" and "Fig. II") and the sections of individual illustrations are labeled A through P so that the decks are easily identified when shown in other perspectives.[98] The expansion of the *Plan's* illustrations allowed the number of slaves depicted to increase from 294 to 482, with the increased figure now closer to the number of slaves allowed under the Dolben Act. The revised text focused on the dimensions of the ship, the amount of space allocated to each slave, a quotation from Falconbridge's *An Account of the Slave Trade on the Coast of Africa*, and the general experiences of the slaves and sailors in the Middle Passage and on the West Coast of Africa.[99] Like the Pennsylvania broadside, the *Description* eliminated the paragraph disavowing emancipation but stopped short of an outright endorsement. At least

8000 copies were produced and distributed in London with additional copies published for audiences in Europe and New England.[100]

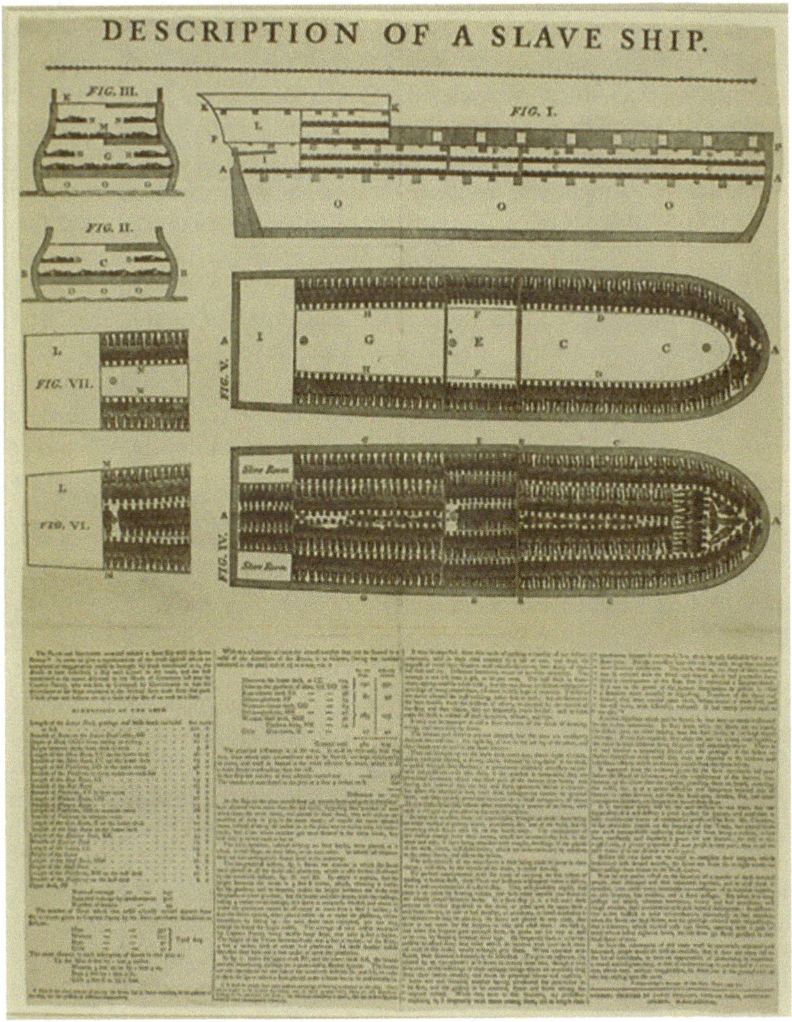

Fig. 3.3 Description of a Slave Ship

Given their unique combination of text and image, the *Plan* and *Description* inaugurated a way of seeing that resonated beyond the slave trade. Abolitionist strategies relied heavily on the racial sentiments of white readers, and this novel visual rhetoric generated equally novel ways of appealing to those sentiments. Clarkson commented in February 1788 that the trade had become "generally known throughout the kingdom," and no less than thirty-five related petitions had been delivered to the House of Commons with several more on their way.[101] His assessment of the trade's notoriety was substantiated by the growing demand for abolitionist paraphernalia. Wedgwood's medallion was by then "highly collectible" and it joined a number of other plates, ceramics, prints, and cameos that demonstrated how thoroughly commercial enterprise was woven into the abolitionist movement.[102] Hence, when the *Plan* and *Description* were published in January and April of 1789, they were received by a public eager for more information about the trade. Most readers had little to no direct experience of the trade itself and as a result, the *Plan* and *Description* came to anchor a "white mythology" through which "the memory of the slave trade became primarily the memory of its glorious abolition."[103] As part of this mythology, the discourse of bad air, the regulation of race and space, and the well-being of sailors and slaves were gathered under the umbrella of abolition and utilized as tools of persuasion. Essentially, the *Plan* and *Description* imply that the act of sharing a racialized environment is a choice and more specifically, a choice that belongs properly to white readers.

Both documents suspend the slaves symbolically and visually in the Middle Passage while removing the ship from its operating milieu. There is nothing to indicate water, wind, light, speed, weather, or sound. This is not a ship in motion; any reference to land or sea is replaced by white, unmarked spaces that channel the viewer's gaze to the images. Detached from its origin and destination and with no indication of voyage length, the hold and its slaves are adrift in an eternal elsewhere. As such, the viewer is insulated not only from the slaves themselves but also from the shared environment of the ship. At the same time, the slaves become by virtue of their generic representation an "abstract and empty vessel vulnerable to the projection of [the viewer's] feelings, ideas, desires, and values."[104] The task of guiding those feelings and ideas falls to the accompanying text, much of which was taken directly from abolitionist essays and the Dolben Act debates. The *Description* contains twenty-two paragraphs to the *Plan's* original eight, with its first ten describing in painstaking detail the dimensions of each deck and "slave room" and the calculations associated with their respective stowage rates. While the two documents touch on the same subjects, the *Description* keeps the reader below decks or bound to the slaves for seventeen paragraphs before switching abruptly to the plight of the sailors.

By underscoring the suffering of the slaves, the narrative conveys to the reader the urgency of the abolitionist cause. To enhance the images' claustrophobic effects, the *Description* includes graphic accounts of the slaves' conditions of confinement. Paragraphs seven and eight report that if the *Brooks* stowed its full complement of 351 captives in its "Men's Room," they would have "only 9 inches each to lay in." Therefore, "instead of lying on their backs" the men were placed "on their sides, or on each other. In which last situation they are not unfrequently found dead in the morning." Paragraph nine informs readers that the height between the deck and the platforms depicted in "fig. II and III" is "no more than 2 feet 7 inches; so that the slaves cannot, when placed either on or under the platform, relieve themselves by sitting up." Paragraphs twelve to sixteen bring the reader briefly on deck to envision the "securing, airing, and exercising" of the slaves. "If the weather is favorable," the text reveals, "they are permitted to remain about one third part of the twenty four hours," with the men "being made to jump in their chains" for exercise. The next paragraph sends the reader back to the hold with a quotation from Falconbridge who recalls finding the slave rooms "so covered with the blood and mucus which had preceded from them in consequence of the flux, that it resembled a slaughter-house." At the end of his account, he passes the narrative from the slaves to the sailors by noting that the slave trade "nearly proved fatal to [him] also."[105] On the level of structure, the text replicates the slaves' voyage by having the reader sit imaginatively in and with the hold's suffocating dimensions before making an appearance on deck. The inadequacy of that visit is punctuated immediately by a reminder of the gruesome conditions to which the slaves return. With less than a quarter of the narrative remaining, the reader is released finally from the seemingly inescapable conditions of the hold.

On the one hand, the text annotates the images in a way that encourages readers to transform the static depictions of the hold into a shared environment. In the process of correlating these images with the experiences of the slaves, they also reconstruct the social-spatial world of the ship. The regulatory functions of the ship's design and organization become tangible as readers are "permitted" to cross between the hold and the deck. Once they arrive at the narrative's final paragraphs, they are asked to take responsibility for the circumstances they helped to create and for which they now serve as witness: "Now let any person reflect on the situation of a number of these devoted people, thus managed and crammed together, and he must think it dreadful…[As] the inhumanity of this trade must be universally admitted and lamented, people would do well to consider, that it does not often fall to the lot of individuals, to have an opportunity of performing so important a moral and religious duty."[106] On the other hand, by relying on readers to animate the immobile and passive figures of the slaves, the broadside reinforces their insulation instead of eroding it. The narrative establishes repeatedly that the slaves' mobility is controlled by the crew and restricted by the ship's built environment. But for the images to convey the depth of their suffering, the reader is required to move in and out of the text and between perspectives of the hold. Put differently, the reader's empathetic identification depends on a way of seeing that distinguishes them fundamentally from the slaves. Crucially, this way of seeing is expanded to include the African environment as paragraphs eighteen to twenty attend to the sailors' health.

In reproducing Clarkson's arguments in *Regulation or Abolition*, the paragraphs forge an association between black suffering and the protection of white bodies. The text cites Africa's "frequent and extremely heavy" rains, the "noisome and infectious effluvia which continually exhale from the slaves," and "boating" as the sailors' primary causes of death. Because the latter was most to blame for exposing them to the African climate, it earned the distinction of being "of a nature offensive and dangerous beyond that of seamen in other services."[107] Like Clarkson's essay, the narrative extends the bad air of the hold and the moral authority of the reader to Africa itself. Whatever social or health risks the slave ship produced were no longer specific to the hold or contained to the ship. For readers, the "moral and religious duty" of abolitionism now included protecting their fellow citizens from dangerous environments. Unlike Clarkson's essay, this new responsibility unfolded in relation to the images of the *Brooks*. The push to protect sailors offered an unspoken reassurance about the end of the slave trade. Even without the slave ship's methodical organization of race and space, it would remain possible to inhabit a racialized environment without being of it or beholden to it. To be sure, the *Description* elicited strong reactions from viewers, and at least publicly, none of them were admiration for the *Brooks*' orderly and efficient arrangement of slaves. And yet, the broadside's "graphic authority" derived substantially from its attention to detail and mathematical precision.[108] To understand how and what that authority contributed to the abolitionist cause, we must first situate the *Description* within a wider historical moment marked by the rise of global capitalism and scientific and mechanical objectivity.

Slave ship design was tied to the development of the plantation system and its mercantile and later capitalist economy. As seventeenth-century plantations grew in number, the increased demand for labor warranted the creation of ships that could transport large numbers of slaves as quickly as possible. Shipbuilding centers dedicated to custom-built slave ships thus sprung up in port cities like Bristol and Liverpool in the mid-eighteenth century, including the shipyard responsible for the *Brooks*. The circulation of shipbuilding technology and ship builders throughout Europe encouraged a greater uniformity in design and construction so that they remained more or less consistent throughout the eighteenth century. So too did the art and science of naval architecture and illustration, which evolved into a formal discipline with the founding of associations like the Society for the Improvement of Naval Architecture in 1791. Subsequently, ship illustrations in the late eighteenth and early nineteenth centuries were more carefully rendered in terms of proportion and included several perspectives. The *Description*'s six new illustrations followed these conventions and as Marcus Wood argues "the conjunction of technical engraving with the depiction of a mass of black human flesh" created a "superb semiotic shock tactic."[109] While naval architecture had a significant impact on the *Plan* and *Description*, their images were also presented in two styles of techno-scientific illustration.

The abolitionist movement was concurrent with an epistemological shift that resulted in the branching of natural history and biology and the strengthening of mechanistic worldviews. Historians of science Lorraine Daston and Peter Galison note that this period was defined by a mode of visual practice called the "truth-to-nature" model.[110] As with any technique of vision, truth-to-nature describes both a (somewhat flexible) code for producing and viewing pictorial work and a set of underlying epistemic values. Primarily concerned with creating idealized images of nature, this model developed in response to the seventeenth-century fashion of illustrating natural objects in all their variation. These earlier illustrations were meant to provide comprehensive surveys for the scientific community and sometimes the more outlandishly the specimens were drawn, the better they were received.[111] By contrast, practitioners operating in the truth-to-nature model opted to "[single] out those features common to the entire species (the *descriptio*) as well as those that differentiated this species from all others in the genus (the *differentia*) but avoided features peculiar to individual member of the species."[112] More importantly, because drawings produced in this manner were perceived as truer to nature's ultimate design, the truth-to-nature model significantly amplified the *Plan* and *Description's* semiotic effects.

To create an image that would reflect the true design of slave ships without emphasizing the "peculiarities" of individual ships, the Plymouth committee used the measurements of a real ship and filled it with generalized representations of slaves.[113] If SEAST had chosen to represent the conditions of an actual voyage of the *Brooks*, they would have left themselves vulnerable to arguments that the ship and voyage were exceptions to an otherwise humane mode of transportation.[114] These tensions between the ideal (i.e., generalized) and the characteristic (i.e., individualized), or between the universal and the particular, are captured in Thomas Cooper's explanation of the abolitionist strategy. In a 1787 letter to the *Manchester Chronicle*, he wrote:

Every man possessed of the common feelings of humanity, and the common principles of morality, even if unacquainted with the particulars of this execrable commerce, mentions it in discourse in terms of disapprobation, and hears it with an ejaculation of abhorrence. But the miseries of five hundred thousand wretches, noticed in general terms, seldom produces a permanent effect among persons, who would shudder at the details of the complicated misery which any individual of the ill-fated group has been doomed to undergo. It is particular distress, and its attendant circumstances, which is calculated to excite compassion.[115]

To his view, the "general terms" of the slave trade were unlikely to rouse the sentiments of viewers but a demonstration of the "particular distress" of "any individual of the ill-fated group" would evoke an enduring compassion. Curiously, in his advocacy for attention to individual experience, Cooper performs the same generalizing gesture he is quick to reject. For if the "particular distress" he seeks belongs technically to "any individual of the ill-fated group," then that distress is no longer "particular." Of course, "particular" does not always mean individualized and in the case of the *Brooks*, "particular" refers to the differences between the "distress" of transportation and other kinds of slavery-related distress.

The *Brooks* may not have been a natural specimen, but what the *Plan* achieved is a truth-to-nature design that rendered the "Slave Ship *Brooks*" taxonomic. The slave ship's *differentia*, or the slave hold and its slaves, separated it from other vessels in the "genus" of transportation ships and thereby from discriminatory evaluations of bad air. Likewise, the hold's dehumanizing optimization of space, or its *descriptio*, became a "species-specific" characteristic of the slave ship. By displaying its *differentia* and *descriptio* in a single image, the *Plan* helped to ensure that the slave ship's conditions of confinement were categorically distinct. Nevertheless, because the London Committee was concerned about representing the slave ship in a manner that avoided any "complaint of exaggeration," they prioritized mathematical and mechanical precision in their improvements to the *Plan*.[116]

The new perspectives added to the *Description* reflected the influence of the Industrial Revolution on techno-scientific representational practice. A drive toward mechanical objectivity emerged at the tail end of the truth-to-nature model, the objective of which was to produce images devoid of any subjective interference. As is the case with new imaging technologies, nineteenth-century inventions like the lithograph and camera lucida inspired alternative ways of perceiving and representing natural organisms. These new visual standards prompted transformations in the scientific self so that interpretive judgment—determinations about which organic structures were common to a species—became secondary to self-discipline. By not resorting to speculation, however well-reasoned, nature could "speak for itself."[117] Consequently, the machine became the exemplar of scientific objectivity because of its supposedly unmediated relationship to the world. "Levers and gears [could] not succumb to [the temptations]" that accompanied free will.[118] Moreover, the "patient, indefatigable, ever-alert machine" could replace or relieve human laborers that were likely to tire or err.[119] Because mechanical objectivity required machines capable of (re)producing images, it did not mature until the mid-nineteenth century when cameras and their prototypes displaced the drawings, engravings, tracings, and etchings of manual reproduction. At the same time, we can see the unmistakable influence of the mechanical age on early nineteenth-century techno-scientific illustrators and the London Committee, the latter of which "employed state-of-the-art drafting techniques commonly used by naval architects to precisely and more fully visualize the [*Brooks*]."[120]

The *Description* utilized naval drafting techniques to make visible the internal logic of slave ship stowage practices. For this reason, it functions also like an exploded-view diagram, or a drawing designed to supplement the human senses by opening a "conceptual space for correlations [not] rooted in direct experience."[121] This graphic form dates to the Renaissance and has remained popular because it "offers just as much intrigue as it does use value."[122] Apart from showing how the parts of a mechanical system are related and assembled, exploded-view drawings invite viewers to "bask in the unfamiliar repleteness" and "configurative nature" of an enigmatic object.[123] In this sense, the *Description* is powerful because it allowed viewers to enjoy the brutal efficiency of race- and slave-making in a way that confirmed rather than compromised their integrity. As an example, in 1789 Clarkson traveled to Paris and spent a year campaigning for the abolition of the French slave trade. While there he met Gabriel Honore de Riqueti Compte de Mirabeau, one of the leading orators of the French assembly and a consummate abolitionist. As Mirabeau prepared to deliver a speech in support of abolition, Clarkson reports that he became "so impressed by [the *Description*] that he ordered a mechanic to make a model of it in wood, at a considerable expense. This model he kept afterwards in his dining room. It was a ship in miniature, about a yard long, and little wooden men and women, which were painted black to represent the slaves, were seen stowed in their proper places."[124] As this account suggests, Mirabeau was moved less by the broadside's narrative appeals and more by its precisely rendered arrangement of slaves. So much so, in fact, that the model's slaves are still "stowed in their proper places."

What the *Description* depicts is a machine that takes the raw materials of space, bodies, and blackness and turns them into slaves. To expose the mechanisms necessary for the ship to function as a machine, the broadside creates a catalogue of parts unavailable to the naked eye. The first illustration of the bow-to-stern longitudinal cross-section (Fig. I) positions the viewer outside of the hold. This illustration sits above the two overhead views of the lower deck (Fig. IV) and platform (Fig. V) and the two overhead views of the quarter decks (Fig. VI and VII). The lateral cross-sections of the lower deck and platform (Fig. II) and the quarter deck (Fig. III) are particularly significant as they contain all five of the *Brooks*' decks.[125] Drawn at a scale of an "eighth of an inch to a foot," they are clearly too shallow to accommodate a standing human.[126] They were designed, in other words, for cargo. The slaves in Figs. II and III are reduced to dense, black lumps and it is only when consulting the overhead views of Figs. IV and V that they become more recognizably human. As an effect, distinguishing humans from objects becomes not only a privilege of perspective but also a pivotal step in the process of race- and slave-making.

Ultimately, the *Plan* and *Description* objectified black bodies in such way that they were rendered objects of a built environment rather than subjects confined within it. The *Plan*'s truth-to-nature model established the precise ordering of race and space as an "organic structure" common to all slave ships. In doing so, the structure became part of nature's "true design" for racialized environments. Then, the *Description*'s diagrammatic properties transformed the trade into a manufacturing operation with the ship serving as a race- and slave-making machine. By virtue of the accompanying narrative, the reader is tasked with deciding not only the fate of the operation but also how to repurpose the machine's "configurative nature" to protect white bodies from dangerous environments. On the one hand, this "uncompromising objectivity" was a fitting representation of the trade's dehumanizing practices.[127] Radical MP Charles Fox alluded to this strategy during the Dolben Act debates when he compared abolitionist arguments to "mathematical propositions."[128] On the other hand, because the *Description* positions readers and sailors as "operators" of the slave ship, their relationship to its environment becomes one of control rather than mutual vulnerability and participation. In this regard, the broadside acts "like a fetish in the classic sense; by focusing on this description of the Middle Passage, of the slave ship, the underlying goal of the abolitionists is obscured."[129] What viewers see when they look at the *Description* is an affirmation of their own humanity, one that depended on the repeated evacuation and resurrection of the humanity of the slaves.

Ghetto Miasma and Bad Air

After the abolition of the slave trade, public attention turned to the plantation system and the related issues of emancipation and segregation. As scholars of race and nature like Katherine McKittrick, Tiffany Lethabo King, and Britt Rusert have shown, the plantation not only played an important role in the development of public health and racial science, but also helped to "[spatialize] early conceptions of urban life within the context of a racial economy."[130] Curiously, despite having similar connections to sanitation and public health, the slave ship rarely appears in the work of nineteenth- and twentieth-century sanitarians and urban reformers. While it remained broadly relevant to discussion of race, empire, and economic stability, it is almost as if its relevance for social reform dissolved with the trade itself. There are, however, two notable exceptions to this dissolution. The first is an 1883 pamphlet by Reverend Andrew Mearns entitled *The Bitter Cry of Outcast London*. Described as one of the most influential texts for British social reform, the pamphlet led directly to the Housing of the Working Classes Act of 1885 which replaced England's laissez-faire approach to housing with one that moved closer to modern city planning.[131] The Act also contains what are now recognized as some of the UK's first public health directives. The pamphlet, which had an "immediate and cataclysmic" effect on the reading public, drew an analogy between the slave hold and living conditions for London's urban poor:

> Few who read these pages have any conception of what these pestilential human rookeries are, where tens of thousands are crowded together amidst horrors which call to mind what we have heard of the middle passage of the slave ship. To get to them you have to penetrate courts reeking with poisonous and malodorous gases arising from accumulations of sewage and refuse scattered in all directions and often flowing beneath your feet; courts, many of them which the sun never penetrates, which are never visited by a breath of fresh air, and which rarely know the virtues of a drop of cleansing water.[132]

By capitalizing on the infamy of the slave ship and its conditions of confinement, Mearns imbued his critique of Victorian social policy with the moral force of abolitionism. If the abolition of the trade represented a moral victory, then the continued existence of overcrowding, poor sanitation, and inadequate ventilation was a sign of moral regression. Correspondingly, the Victorian middle class reacted most strongly to his argument that such conditions lead inevitably to crime: "That people condemned to exist under such conditions take to drink and fall into sin is surely a matter for little surprise…There can be no question that numbers of habitual criminals would never have become such, had they not by force of circumstances been packed together in these slums with those who were hardened in crime."[133] Under the terms of his analogy, moral degradation and criminality were not only effects of poverty, but were also emergent properties of the slave hold itself. Seventy-five years after the abolition of the trade, the slave ship environment became a standard against which moral progress and public health could be measured. This standardization was in fact repeated in the second notable exception, Edward Bellamy's 1888 utopian time traveling novel *Looking Backward*.

As a frank commentary on socialism, capitalism, and urban reform, *Looking Backward* inspired political and social movements both at the time of its publication and since. Most importantly for this chapter, the novel was also a major source of inspiration for Ebenezer Howard's Garden City movement, which "proposed to solve, or at least ameliorate, the problem of the Victorian city by exporting a goodly proportion of its people and its jobs to new, self-contained, constellations of new towns built in open countryside."[134] Like the novel itself, the Garden City movement was hugely influential and in large part because it laid the groundwork for modern urban planning and the practice of zoning. In the final chapter of *Looking Backward*, protagonist Julian West dreams of returning to the Boston of 1887 and compares its tenement district to the hold of a slave ship:

> From the black doorways and windows of the rookeries on every side came gusts of fetid air. The streets and alleys reeked with the effluvia of a slave ship's between-decks. As I passed I had glimpses within of pale babies gasping out their lives amid sultry stenches, of hopeless-faced women deformed by hardship, retaining of womanhood no trait save weakness, while from the windows leered girls with brows of brass.[135]

Like Mearns' pamphlet, the novel recruits the slave ship to establish a timeline of moral progress. In this "scene of squalor," the ship's "effluvia" marks the poverty and hardship of industrial capitalism with the stench of the past.[136] The way to avoid further regression, therefore, is to ensure that white bodies are protected from such dangerous environments. Instead of drawing its power from the suffering of the slaves, the novel's analogy turns the slave ship into a cautionary symbol against white moral degradation.

While the pamphlet and novel are cited often in relation to public health and urban planning, their references to the slave ship are left unremarked. And yet, I submit that it is not coincidental that the slave ship appears in two of the most important texts for these fields. Nor is it coincidental that these appearances reinforce a correlation between blackness, bad air, and the built environment. In treating the slave ship's bad air as a threat to white well-being, the pamphlet and novel not only (re)racialize its conditions of confinement but also obfuscate how the effects of that racialization accumulate over time in the built, natural, and social environments. Clearly, poor air quality, overcrowding, and inadequate sanitation were and are still treated as public health concerns, including in "densely populated urban immigrant neighborhoods."[137] But as Epstein points out, we still lack a satisfying explanation for why "white people who live in cities with large black populations have higher death rates than whites with the same income who live in cities with smaller black populations." She continues, "it made me wonder whether the deprived, polluted, roach-infested, stressful conditions in which poor blacks live aren't affecting all of us, to some degree."[138] Indeed, if antiblackness is a cumulative environmental effect, then "all of us" share in its consequences even or perhaps especially when we choose not to acknowledge it as such.

Notes

1. Helen Epstein, "Ghetto Miasma–Enough to Make You Sick?" *The New York Times Magazine*, October 12, 2003, https://www.nytimes.com/2003/10/12/magazine/ghetto-miasma-enough-to-make-you-sick.html.
2. Epstein, "Ghetto Miasma."
3. Bernard L. Bloom, "The 'Medical Model,' Miasma Theory, and Community Mental Health," *Community Mental Health Journal* 1 (1965): 333–338; Jan Paul Vandenbroucke, "Is 'The Causes of Cancer' a Miasma Theory for the End of the Twentieth Century?," *International Journal of Epidemiology* 17, no. 4 (1988): 708–709; and Melanie A. Kiechle, "Revisiting a 19th Century Medical Idea Could Help Address Covid-19," *Washington Post*, April 21, 2021, https://www.washingtonpost.com/outlook/2021/04/21/revisiting-19th-century-medical-idea-could-help-address-covid-19/. Medical practitioners and historians have also suggested that miasma theory can help us to better understand community mental health and even sanitation practices in the wake of the Covid-19 pandemic.
4. Julie Sze, *Noxious New York: The Racial Politics of Urban Health and Environmental Justice* (MIT Press, 2006), 30; Linda Nash, *Inescapable Ecologies: A History of Environment, Disease, and*

Knowledge (University of California Press, 2006), 190–191; and Max Liboiron, "Plasticizers: A Twenty-First-Century Miasma," in *Accumulation*, eds. Jennifer Gabrys, Gay Hawkins, and Mike Michael (Routledge, 2013), 148–163.

5. Liboiron, "Plasticizers," 134.
6. Liboiron, "Plasticizers," 134.
7. Liboiron, "Plasticizers," 135–136.
8. Sze, *Noxious New York*, 102–103.
9. Sze, *Noxious New York*, 93.
10. "Cumulative Impacts Research," United Stated Environmental Protection Agency, last updated on February 19, 2025, https://www.epa.gov/healthresearch/cumulative-impacts-research.
11. Christina Sharpe, *In the Wake: On Blackness and Being* (Duke University Press, 2016), 106 (emphasis in the original).
12. Sharpe, *Wake*, 111.
13. Luke W. Cole and Sheila R. Foster, *From the Ground Up: Environmental Racism and the Rise of the Environmental Justice Movement* (NYU Press, 2001), 63; Sze, *Noxious New York*, 91–108.
14. Gabriel O. Apata, "'I Can't Breathe': The Suffocating Nature of Racism," *Theory, Culture & Society* 37, no. 7–8 (2020): 242.
15. Apata, "I Can't Breathe," 246.
16. Ali Winston, "Medical Examiner Testifies Eric Garner Died of Asthma Caused by Officer's Chokehold," *The New York Times*, May 15, 2019, https://www.nytimes.com/2019/05/15/nyregion/eric-garner-death-daniel-pantaleo-chokehold.html; Sheri Fink, "Floyd's Cause of Death Is Crucial in Chauvin's Trial. Forensic Pathologists Explain," *The New York Times*, April 8, 2021, https://www.nytimes.com/2021/04/08/us/george-floyd-cause-of-death.html.
17. Daniel B. Domingues da Silva, "'I Can't Breathe' is Tied to a Long History of Black Asphyxiation," *The Washington Post*, August 14, 2020, https://www.washingtonpost.com/outlook/2020/08/14/i-cant-breathe-is-tied-long-history-black-asphyxiation/; Ki'Amber Thompson, "Toward a World Where We Can Breathe: Abolitionist Environmental Justice Praxis," *Annals of the American Association of Geographers* 113, no.7 (2023): 1699–1710.
18. Apata, "I Can't Breathe," 251.

19. John Duffy, *The Sanitarians: A History of American Public Health*, (University of Illinois Press, 1992); Hsuan L. Hsu, *The Smell of Risk: Environmental Disparities and Olfactory Aesthetics* (New York University Press, 2020); Anthony Kessel, *Air, the Environment and Public Health*, (Cambridge University Press, 2006); Russell P. Lopez, *The Built Environment and Public Health* (John Wiley & Sons, 2012); and George Rosen, *A History of Public Health*. (Johns Hopkins University Press, 2015).
20. Denise L. Lawrence and Setha M. Low, "The Built Environment and Spatial Form," *Annual Review of Anthropology* 19, no. 1 (1990): 453–505; Sebastian Moffatt and Niklaus Kohler, "Conceptualizing the Built Environment as a Social–Ecological System," *Building Research & Information* 36, no. 3 (2008): 248–268. The term "built environment" began appearing in planning literature in the late 1960's and was in common use by the mid-1970s.
21. Caroline Imani Collins and K. Wayne Yang, "Between Ship and Sea and Shore: Spatial Practices from the Black Pacific," *Scroope* 33, no.1 (2024): 22–45.
22. Cheryl Finley, *Committed to Memory: The Slave Ship Icon in the Black Atlantic Imagination* (Princeton University Press, 2018); Marcus Wood, *Blind Memory* (Routledge, 2000); and Jane Webster, "The Unredeemed Object: Displaying Abolitionist Artifacts in 2007," *Slavery & Abolition: A Journal of Slave and Post-Slave Studies* 30, no. 2 (2009): 311–325. These notable exceptions examine the *Brooks'* influence on contemporary visual culture, nineteenth- and twentieth-century art, and/or the histories of naval architecture and illustration.
23. Richard H Steckel and Richard A. Jensen, "New Evidence on the Causes of Slave and Crew Mortality in the Atlantic Slave Trade," *The Journal of Economic History* 46, no. 1 (1986): 47.
24. Stephen Halliday, "Death and Miasma in Victorian London: An Obstinate Belief," *British Medical Journal* 323, no. 7327 (2001): 469–1471.
25. Halliday, "Death and Miasma," 1469.
26. Thomas Carlyle, *Chartism* (Chapman and Hall, 1842).
27. Christopher Hamlin, *Public Health and Social Justice in the Age of Chadwick: Britain, 1800–1854* (Cambridge University Press, 1998), 13–15.

28. Hamlin, *Public Health*, 12 (emphasis in the original).
29. Hamlin, *Public Health*, 165.
30. Carl A. Zimring, *Clean and White* (New York University Press, 2016), 32–34; 84–86.
31. Stephen C. Kenny, "Capturing Racial Pathology: American Medical Photography in the Era of Jim Crow," *American Journal of Public Health* 110, no. 1 (2020): 75–83.
32. Sze, *Noxious New York*, 36–37; Zimring, *Clean and White*, 3.
33. Zimring, *Clean and White*, 3–4.
34. Garret Power, "Apartheid Baltimore Style: The Residential Segregation Ordinance of 1910–1913," *Maryland Law Review* 42 (1983): 301.
35. Zimring, *Clean and White*, 82; 98.
36. Robin Haines and Ralph Shlomowitz, "Explaining the Mortality Decline in the Eighteenth-Century British Slave Trade," *Economic History Review* 53, no. 2 (2000): 262.
37. David Boyd Haycock, "Health, Medicine and the Maritime World: A History of Two Centuries," in *Health and Medicine at Sea, 1700–1900*, eds. David Boyd Haycock and Sally Archer (Boydell & Brewer, 2009), 8–10.
38. Haines and Shlomowitz, "Mortality Decline," 279.
39. Aaron S. Fogleman, "From Slaves, Convicts, and Servants to Free Passengers: The Transformation of Immigration in the Era of the American Revolution," *The Journal of American History* 85, no. 1 (1998): 44; Christopher Tomlins, "Reconsidering Indentured Servitude: European Migration and the Early American Labor Force, 1600–1775," *Labor History* 42, no. 1 (2001): 8–9; and "Slavery and the British Transatlantic Slave Trade," The National Archives of the United Kingdom, accessed July 17, 2025, https://www.nationalarchives.gov.uk/help-with-your-research/research-guides/british-transatlantic-slave-trade.
40. Sowande M. Mustakeem, *Slavery at Sea: Terror, Sex, and Sickness in the Middle Passage* (University of Illinois Press, 2016); Brian Vale and Griffith Edwards, *Physician to the Fleet: The Life and Times of Thomas Trotter, 1760–1832*, (Boydell & Brewer, 2011), 28–33; and Haines and Shlomowitz, "Mortality Decline," 274.
41. James Lind, *An Essay on the Most Effectual Means of Preserving the Health of Seamen in the Royal Navy* (D. Wilson, 1762), 3.

42. Kevin Siena, *Rotten Bodies: Class and Contagion in Eighteenth-Century Britain* (Yale University Press, 2019), 13.
43. Haines and Shlomowitz, "Mortality Decline," 277.
44. John Pringle, *Observations on the Nature and Cure of Hospital and Jail Fevers* (A. Millar and D. Wilson, 1750), 4 - 8.
45. John Pringle, *Observations on the Diseases of the Army, in Camp and Garrison: In Three Parts: with an Appendix, Containing Some Papers of Experiments, Read at Several Meetings of the Royal Society* (A. Millar, D. Wilson, T. Durham and T. Payne, 1753), 40.
46. Pringle, *Jail Fever*, 11; Pringle, *Army*, 26.
47. Siena, *Rotten Bodies*, 109.
48. Suman Seth, *Difference and Disease: Medicine, Race, and the Eighteenth-Century British Empire* (Cambridge University Press, 2018).
49. Seth, *Difference and Disease*, 266.
50. Edward Long, *The History of Jamaica* vol. 2 (T. Lowndes, 1774), 412.
51. Seth, *Difference and Disease*, 265.
52. Seth, *Difference and Disease*, 112–163; 266–276.
53. Andrew Wilson, *Rational Advice to the Military, When Exposed to the Inclemency of Hot Climates and Seasons* (W. Richardson, 1780), 270.
54. Richard B. Sheridan, "The Guinea Surgeons on the Middle Passage: The Provision of Medical Services in the British Slave Trade," *The International Journal of African Historical Studies* 14, no. 4 (1981): 610.
55. Thomas Aubrey, *The Sea-Surgeon, or the Guinea Man's Vade Mecum* (John Clarke, 1729), 121.
56. Haines and Shlomowitz, "Mortality Decline," 272; Pringle, *Army*, xii, 110, 401.
57. Haines and Shlomowitz, "Mortality Decline," 280.
58. Herbert S. Klein, Stanley L. Engerman, Robin Haines, and Ralph Shlomowitz, "Transoceanic Mortality: The Slave Trade in Comparative Perspective," *The William and Mary Quarterly* 58, no. 1 (2001): 93–118; and Eric Herschthal, "Slavery, Health, and Healing Now: The State of the Field," *Journal of the History of Medicine and Allied Sciences* 77, no. 1 (2022): 1–23. There is considerable debate among historians of the slave trade as

to the causes of mortality on British slave ships. While some historians believe overcrowding was a significant factor in slave deaths, others suggest its impact on health was either negligible or counteracted by the adoption of sanitary procedures.

59. Seth, *Difference and Disease*, 19. See also Seymour Drescher, "The Ending of the Slave Trade and the Evolution of European Scientific Racism," *Social Science History* 14, no. 3 (1990): 415–450. While the British abolitionist movement was not formalized until 1787, religious and political leaders began to question publicly the immorality of slavery in the mid-eighteenth century.
60. Jim Downs, *Maladies of Empire* (Harvard University Press, 2020); and Seth, *Difference and Disease*, 241–290.
61. Christopher M. Blakley, "Ship Fever, Confinement, and the Racialization of Disease," *Studies in History and Philosophy of Science* 95 (2022): 97.
62. Charles Bisset, *An Essay on the Medical Constitution of Great Britain* (A. Millar, 1762); and Charles Bisset, *Candid and Satisfactory Answers to the Several Criticisms of the Critical Reviewers, on an essay on the Medical Constitution of Great Britain* (A. Millar, and D. Wilson, 1763), 11, 14.
63. Lind, *Health of Seamen*, 3, 95; James Lind, *An Essay on Diseases Incidental to Europeans in Hot Climates* (T. Becket and P. A. de Hondt, 1768), 134; Blakley "Ship Fever," 100.
64. James Veitch, *A Letter to the Commissioners for Transports, and Sick and Wounded Seamen, on the Non-Contagious Nature of the Yellow Fever* (T. and G. Underwood, 1818), 58–64.
65. Blakley, "Ship Fever," 102.
66. Seth, *Difference and Disease*, 266.
67. Alexander Falconbridge, *An Account of the Slave Trade On the Coast of Africa* (J. Phillips, 1788), 28–29.
68. John Newton, *Thoughts Upon the African Slave Trade* (J. Buckley, 1788), 34.
69. Thomas Trotter, *House of Commons Sectional Papers of the Eighteenth Century* (Wilmington, 1875), LXXIII, 81–101, quoted in Vale and Edwards, *Physician to the Fleet*, 62.
70. Sharpe, *Wake*, 108–113; See also Anna More, "The Early Portuguese Slave Ship and the Infrastructure of Racial Capitalism," *Social Text* 40, no. 4 (2022): 17–41.

71. Vale and Edwards, *Physician to the Fleet*, 48–68; Sowande M. Mustakeem, "'She Must Go Overboard & Shall Go Overboard": Diseased Bodies and the Spectacle of Murder at Sea," *Atlantic Studies* 8, no. 3 (2011): 301–316; and Marcus Rediker, *The Slave Ship: A Human History* (Penguin Books, 2007).
72. Rediker, *Slave Ship*, 68–71; and Arnold Zuckerman, "Scurvy and the Ventilation of Ships in the Royal Navy: Samuel Sutton's Contribution," *Eighteenth-Century Studies* 10, no. 2 (Winter 1976–1977): 228.
73. Hamish Maxwell-Stewart and Ralph Shlomowitz, "Mortality and Migration: A Survey," in *Health and Medicine at Sea, 1700–1900*, eds. David Boyd Haycock and Sally Archer (Boydell & Brewer, 2009), 128–142. Slave ship mortality rates remained much higher than those of other kinds of vessels throughout the entirely of the trade.
74. Thomas Clarkson, *The History of the Rise, Progress, and Accomplishment of the Abolition of the African-Slave Trade by the British Parliament*, vol. 1 (L. Taylor, 1808), 520.
75. Clarkson, *History*, 1:520.
76. Clarkson, *History*, 1:520; and James W. LoGerfo, "Sir William Dolben and 'The Cause of Humanity': The Passage of the Slave Trade Regulation Act of 1788," *Eighteenth-Century Studies* 6, no. 4 (1973): 445.
77. LoGerfo, "William Dolben," 446.
78. Testimony of Robert Norris, *House of Commons Sessional Papers*, 73:4–5, 8, 10; 69:203, quoted in Rediker, *Slave Ship*, 328.
79. "Minutes of the Evidence Taken in the Last Session of Parliament before the Committee of the Whole House, to Whom the Bill for Providing Certain Temporary Regulations Respecting the Transportation of the Natives of Africa, in British Ships, to the West Indies, and Elsewhere Was Committed," *House of Commons Sessional Papers* 68: 4–6; quoted in Paul E. Sampson, "'The Lungs of a Ship': Ventilation, Acclimatization, and Labor in the Maritime Environment, 1740–1800," *History of Science* 61, no. 2 (2023): 231.
80. Thomas Clarkson, *The History of the Rise, Progress, and Accomplishment of the Abolition of the African-Slave Trade by the British Parliament*, vol. 2 (L. Taylor, 1808), 69.
81. Clarkson, *History*, 2:60.

82. Clarkson, *History*, 2:61, 339.
83. Sampson, "Lungs," 232.
84. Thomas Clarkson, *An Essay on the Comparative Efficiency of Regulation or Abolition, as Applied to the Slave Trade: Shewing That the Latter Only Can Remove the Evils to be Found in That Commerce* (J. Phillips, 1789), 51, 62.
85. Clarkson, *Regulation or Abolition*, 62.
86. Clarkson, *Regulation or Abolition*, 43, 45.
87. Clarkson, *Regulation or Abolition*, 45; Thomas Clarkson, *An Essay on the Impolicy of the African Slave Trade. In Two Parts* (J. Phillips, 1788), 53.
88. Clarkson, *History*, 2:111.
89. Slave Trade Act 1789, 29 Geo. 3, c.66 (Eng.).
90. Sampson, "Lungs," 235.
91. Rediker, *Slave Ship*, 310. The total number of slaves carried by the *Brooks* varies according to source, as does the number of slaves who survived their voyages.
92. Clarkson, *History*, 2:111.
93. Finley, *Committed to Memory*, 72–74; Clarkson, *History*, 2:115.
94. Clarkson, *History*, 1:520.
95. Finley, *Committed to Memory*, 36.
96. Sir William Elford, *Plan of an African Ship's Lower Deck with Negroes in the Proportion of Only One to a Ton* (Trewman and Haydon, 1789).
97. *Remarks on the Slave Trade* (Matthew Carey for the Pennsylvania Society for Promoting the Abolition of Slavery, 1789).
98. Rediker, *Slave Ship*, 331. A later edition of the Description added an eighth illustration that depicted a longitudinal view of a stylized slave ship in the mid of a slave rebellion. That image, entitled "Representation of an Insurrection on board a Slave-Ship," was likely added after the Haitian Rebellion became a part of the larger debates over slavery and the slave trade.
99. Rediker, *Slave Ship*, 317.
100. Marcus, *Blind Memory*; 17–18; J.R. Oldfield, *Popular Politics and British Anti-Slavery* (Manchester University Press, 1995), 163–166.
101. Clarkson, *History*, 1:469.

102. Oldfield, *Popular Politics*, 155–179. See also Charlotte Sussman, *Consuming Anxieties: Consumer Protest, Gender, and British Slavery* (Stanford University Press, 2000).
103. Wood, *Blind Memory*, 8.
104. Saidiya Hartman, *Scenes of Subjection* (Oxford University Press, 1997), 21.
105. *Description of a Slave Ship* (James A. Phillips for the London Committee of the Society for Effecting the Abolition of the Slave Trade, 1789).
106. *Description of a Slave Ship*.
107. *Description of a Slave Ship*.
108. Wood, *Blind Memory*, 25.
109. Wood, *Blind Memory*, 26–27.
110. Lorraine Daston and Peter Galison, *Objectivity* (Zone Books, 2007).
111. Francis Bacon, *Novum Organum* (Clarendon Press, 1878).
112. Daston and Galison, *Objectivity*, 60.
113. Clarkson, *History*, 1:535.
114. Rediker, *Slave Ship*; Nicholas Radburn and David Eltis, "Visualizing the Middle Passage: The Brooks and the Reality of Ship Crowding in the Transatlantic Slave Trade," *Journal of Interdisciplinary History* 49, no. 4 (2019): 533–537. Ironically, the *Brooks* often carried more slaves than were allowed under the Dolben Act, so the calculations that determined the number of slaves portrayed in the *Description* likely underestimated the number of slaves carried on a typical voyage.
115. Thomas Cooper, *Letters on the Slave Trade: First Published in Wheeler's Manchester Chronicle; and Since Re-Printed with Additions and Alterations* (C. Wheeler, 1787).
116. Clarkson, *History*, 1:535.
117. Daston and Galison, *Objectivity*, 120.
118. Daston and Galison, *Objectivity*, 123.
119. Daston and Galison, *Objectivity*, 123.
120. Finley, *Committed to Memory*, 58. See also David McGee, "From Craftsmanship to Draftsmanship: Naval Architecture and the Three Traditions of Early Modern Design," *Technology and Culture* 40, no. 2 (1999): 209–236. Naval illustrations helped to establish the ship as a world- and empire-making machine

in part by adopting new mathematical models for calculating displacement, speed, and buoyancy.

121. John Bender and Michael Marrinan, *The Culture of Diagram* (Stanford University Press, 2010), 17.
122. Ian Bogost, *Alien Phenomenology* (University of Minnesota Press, 2012), 51.
123. Bogost, *Alien Phenomenology*, 51–52.
124. Clarkson, *History*, 2:153.
125. Finley, *Committed to Memory*, 58–59.
126. *Description of a Slave Ship.*
127. Wood, *Blind Memory*, 23.
128. Clarkson, *History*, 2:11, 1:545.
129. Finley, *Committed to Memory*, 41–42.
130. Katherine McKittrick, "Plantation Futures," *small axe* 17, no. 3 (2013): 8. See also Katherine McKittrick, *Demonic Grounds: Black Women and the Cartographies of Struggle* (University of Minnesota Press, 2006); Tiffany Lethabo King, *The Black Shoals: Offshore Formations of Black and Native Studies* (Duke University Press, 2019); Britt Rusert, *Fugitive Science: Empiricism and Freedom in Early African American Culture*, vol. 10 (NYU Press, 2017); and Britt Rusert, "Plantation Ecologies: The Experimental Plantation in and against James Grainger's 'The Sugar-Cane,'" *Early American Studies* (2015): 341–373.
131. Peter Hall, *Cities of Tomorrow: An Intellectual History of Urban Planning and Design Since 1880* (John Wiley & Sons, 2014), 13–19.
132. Anthony S. Wohl, "The Bitter Cry of Outcast London," *International Review of Social History* 13, no. 2 (1968): 189; Andrew Mearns, *The Bitter Cry of Outcast London: An Inquiry into the Condition of the Abject Poor* (London: James Clarke, 1883), 7.
133. Hall, *Cities of Tomorrow*, 16; Mearns, *Bitter Cry*, 11.
134. Hall, *Cities of Tomorrow*, 7; 95–96.
135. Edward Bellamy, *Looking Backward, 2000–1887*, ed. Matthew Beaumont (Oxford University Press, 2002), 188–189.
136. Bellamy, *Looking Backward*, 188.
137. Sze, *Noxious New York*, 42.
138. Epstein, "Ghetto Miasma."

References

Apata, Gabriel O. "'I Can't Breathe': The Suffocating Nature of Racism." *Theory, Culture & Society* 37, no. 7–8 (2020): 241–254.

Aubrey, Thomas. *The Sea-Surgeon, or the Guinea Man's Vade Mecum*. John Clarke, 1729.

Bacon, Francis. *Novum Organum*. Clarendon Press, 1878.

Bellamy, Edward. *Equality*. D. Appleton, 1897.

Bellamy, Edward. *Looking Backward, 2000–1887*. Edited by Matthew Beaumont. Oxford World's Classics. Oxford University Press, 2002.

Bender, John and Michael Marrinan. *The Culture of Diagram*. Stanford University Press, 2010.

Bisset, Charles. *Candid and Satisfactory Answers to the Several Criticisms of the Critical Reviewers, on an essay on the Medical Constitution of Great Britain*. A. Millar and D. Wilson, 1763.

Bisset, Charles. *An Essay on the Medical Constitution of Great Britain*. A. Millar, 1762.

Blakley, Christopher M. "Ship Fever, Confinement, and the Racialization of Disease." *Studies in History and Philosophy of Science* 95 (2022): 96–103.

Bloom, Bernard L. "The 'Medical Model,' Miasma Theory, and Community Mental Health." *Community Mental Health Journal* 1 (1965): 333–338.

Bogost, Ian. *Alien Phenomenology*. University of Minnesota Press, 2012.

Chadwick, Edwin. *Report on the Sanitary Condition of the Labouring Population of Great Britain*. W. Clowes, 1842.

Clarkson, Thomas. *An Essay on the Comparative Efficiency of Regulation or Abolition, as Applied to the Slave Trade: Shewing That the Latter Only Can Remove the Evils to be Found in That Commerce*. J. Phillips, 1789.

Clarkson, Thomas. *An Essay on the Impolicy of the African Slave Trade. In Two Parts*. J. Phillips, 1788.

Clarkson, Thomas. *The History of the Rise, Progress, and Accomplishment of the Abolition of the African-Slave Trade by the British Parliament*, vol. 1 and 2 London, England: L. Taylor, 1808.

Cole, Luke W. and Sheila R. Foster, *From the Ground Up: Environmental Racism and the Rise of the Environmental Justice Movement*. NYU Press, 2001.

Collins, Caroline Imani, and K. Wayne Yang. "Between Ship and Sea and Shore: Spatial Practices from the Black Pacific." *Scroope* 33, no.1 (2024): 22–45.

Cooper, Thomas. *Letters on the Slave Trade: First Published in Wheeler's Manchester Chronicle; and Since Re-Printed with Additions and Alterations*. C. Wheeler, 1787.

da Silva, Daniel B. Domingues. "'I Can't Breathe' is Tied to a Long History of Black Asphyxiation." *The Washington Post*, August 14, 2020. https://www.washingtonpost.com/outlook/2020/08/14/I-cant-breathe-is-tied-long-history-black-asphyxiation/.

Daston, Lorraine and Peter Galison. *Objectivity*. Zone Books, 2007.
Description of a Slave Ship. James A. Phillips for the London Committee of the Society for Effecting the Abolition of the Slave Trade, 1789.
Drescher, Seymour. "The Ending of the Slave Trade and the Evolution of European Scientific Racism." *Social Science History* 14, no. 3 (1990): 415–450.
Downs, Jim. *Maladies of Empire*. Harvard University Press, 2020.
Duffy, John. *The Sanitarians: A History of American Public Health*. University of Illinois Press, 1992.
Elford, Sir William. *Plan of an African Ship's Lower Deck with Negroes in the Proportion of Only One to a Ton*. Trewman and Haydon, 1789.
Epstein, Helen. "Ghetto Miasma–Enough to Make You Sick?" *The New York Times Magazine*, October 12, 2003. https://www.nytimes.com/2003/10/12/magazine/ghetto-miasma-enough-to-make-you-sick.html.
Falconbridge, Alexander. *An Account of the Slave Trade on the Coast of Africa*. J. Phillips, 1788.
Fink, Sheri. "Floyd's Cause of Death Is Crucial in Chauvin's Trial. Forensic Pathologists Explain." *The New York Times*, April 8, 2021. https://www.nytimes.com/2021/04/08/us/george-floyd-cause-of-death.html.
Finley, Cheryl. *Committed to Memory: The Slave Ship Icon in the Black Atlantic Imagination*. Princeton University Press, 2018.
Fogleman, Aaron S. "From Slaves, Convicts, and Servants to Free Passengers: The Transformation of Immigration in the Era of the American Revolution." *The Journal of American History* 85, no. 1 (1998): 43–76.
Francis, Jacqueline. "The Brooks Slave Ship Icon: A 'Universal Symbol'?" *Slavery & Abolition: A Journal of Slave and Post-Slave Studies* 30, no. 2 (2009): 327–338.
Haines, Robin, and Ralph Shlomowitz. "Explaining the Mortality Decline in the Eighteenth-Century British Slave Trade." *Economic History Review* 53, no. 2 (2000): 262–283.
Hall, Peter. *Cities of Tomorrow: An Intellectual History of Urban Planning and Design Since 1880*. John Wiley & Sons, 2014.
Halliday, Stephen. "Death and Miasma in Victorian London: An Obstinate Belief." *British Medical Journal* 323, no. 7327 (2001): 1469–1471.
Hamlin, Christopher. *Public health and social justice in the age of Chadwick: Britain, 1800–1854*. Cambridge University Press, 1998.
Hartman, Saidiya. *Scenes of Subjection*. Oxford University Press, 1997.
Haycock, David Boyd. "Health, Medicine and the Maritime World: A History of Two Centuries." In *Health and Medicine at Sea, 1700–1900*, edited by David Boyd Haycock and Sally Archer. Boydell & Brewer, 2009.
Herschthal, Eric. "Slavery, Health, and Healing Now: The State of the Field." *Journal of the History of Medicine and Allied Sciences* 77, no. 1 (2022): 1–23.

Hsu, Hsuan L. *The Smell of Risk: Environmental Disparities and Olfactory Aesthetics*. New York University Press, 2020.

Kenny, Stephen C. "Capturing Racial Pathology: American Medical Photography in the Era of Jim Crow." *American Journal of Public Health* 110, no. 1 (2020): 75–83.

Kessel, Anthony. *Air, the Environment and Public Health*. Cambridge University Press, 2006.

Kiechle, Melanie. "Revisiting a 19th Century Medical Idea Could Help Address Covid 19." *Washington Post*, April 21, 2001. https://www.washingtonpost.com/outlook/2021/04/21/revisiting-19th-century-medical-idea-could-help-address-covid-19/.

King, Tiffany Lethabo. *The Black Shoals: Offshore Formations of Black and Native Studies*. Duke University Press, 2019.

Klein, Herbert S., Stanley L. Engerman, Robin Haines, and Ralph Shlomowitz. "Transoceanic Mortality: The Slave Trade in Comparative Perspective." *The William and Mary Quarterly* 58, no. 1 (2001): 93–118.

Lawrence, Denise L., and Setha M. Low. "The Built Environment and Spatial Form." *Annual Review of Anthropology* 19, no. 1 (1990): 453–505.

Liboiron, Max. "Plasticizers: A Twenty-First-Century Miasma." In *Accumulation*, edited by Jennifer Gabrys, Gay Hawkins, and Mike Michael. Routledge, 2013.

Lind, James. *An Essay on Diseases Incidental to Europeans in Hot Climates* T. Becket and P. A. de Hondt, 1768.

Lind, James. *An Essay on the Most Effectual Means of Preserving the Health of Seamen in the Royal Navy*. D. Wilson, 1762.

LoGerfo, James W. "Sir William Dolben and 'The Cause of Humanity': The Passage of the Slave Trade Regulation Act of 1788." *Eighteenth-Century Studies* 6, no. 4 (1973): 431–451.

Long, Edward. *The History of Jamaica* vol. 2. T. Lowndes, 1774.

Lopez, Russell P. *The Built Environment and Public Health*. John Wiley & Sons, 2012.

Maxwell-Stewart, Hamish, and Ralph Shlomowitz. "Mortality and Migration: A Survey." In *Health and Medicine at Sea, 1700–1900*, edited by David Boyd Haycock and Sally Archer. Boydell & Brewer, 2009.

McGee, David. "From Craftsmanship to Draftsmanship: Naval Architecture and the Three Traditions of Early Modern Design." *Technology and Culture* 40, no. 2 (1999): 209–236.

McKittrick, Katherine. *Demonic Grounds: Black Women and the Cartographies of Struggle*. University of Minnesota Press, 2006.

McKittrick, Katherine. "Plantation Futures." *small axe* 17, no. 3 (2013): 1–15.

Mearns, Andrew. *The Bitter Cry of Outcast London: An Inquiry into the Condition of the Abject Poor*. London: James Clarke, 1883.

Moffatt, Sebastian, and Niklaus Kohler. "Conceptualizing the Built Environment as a Social–Ecological System." *Building Research & Information* 36, no. 3 (2008): 248–268.

More, Anna. "The Early Portuguese Slave Ship and the Infrastructure of Racial Capitalism." *Social Text* 40, no. 4 (2022): 17–41.

Mustakeem, Sowande M. "'She Must Go Overboard & Shall Go Overboard": Diseased Bodies and the Spectacle of Murder at Sea." *Atlantic Studies* 8, no. 3 (2011): 301–316.

Mustakeem, Sowande M. *Slavery at Sea: Terror, Sex, and Sickness in the Middle Passage*. University of Illinois Press, 2016.

Nash, Linda Lorraine. *Inescapable Ecologies: A History of Environment, Disease, and Knowledge*. University of California Press, 2006.

The National Archives of the United Kingdom. "Slavery and the British Transatlantic Slave Trade." Accessed July 17, 2025. https://www.nationalarchives.gov.uk/help-with-your-research/research-guides/british-transatlantic-slave-trade-records/.

Newton, John. *Thoughts Upon the African Slave Trade*. J. Buckley, 1788.

Oldfield, J.R. *Popular Politics and British Anti-Slavery*. Manchester University Press 1995.

Power, Garret. "Apartheid Baltimore Style: The Residential Segregation Ordinance of 1910–1913." *Maryland Law Review* 42 (1983): 296–301.

Pringle, John. *Observations on the Diseases of the Army, in Camp and Garrison: In Three Parts: with an Appendix, Containing Some Papers of Experiments, Read at Several Meetings of the Royal Society*. A. Millar, D. Wilson and T. Durham and T. Payne, 1753.

Pringle, John. *Observations on the Nature and Cure of Hospital and Jail Fevers*. A. Millar, and D. Wilson, 1750.

Radburn, Nicholas, and David Eltis. "Visualizing the Middle Passage: The Brooks and the Reality of Ship Crowding in the Transatlantic Slave Trade." *Journal of Interdisciplinary History* 49, no. 4 (2019): 533–565.

Rediker, Marcus. *The Slave Ship: A Human History*. Penguin Books, 2007.

Remarks on the Slave Trade. Matthew Carey for the Pennsylvania Society for Promoting the Abolition of Slavery, 1789.

Rusert, Britt. *Fugitive Science: Empiricism and Freedom in Early African American Culture*. New York University Press, 2017.

Rusert, Britt. "Plantation Ecologies: The Experimental Plantation in and against James Grainger's 'The Sugar-Cane.'" *Early American Studies* (2015): 341–373.

Sampson, Paul E. "'The Lungs of a Ship': Ventilation, Acclimatization, and Labor in the Maritime Environment, 1740–1800." *History of Science* 61, no. 2 (2023): 214–235.

Seth, Suman. *Difference and Disease: Medicine, Race, and the Eighteenth-Century British Empire*. Cambridge University Press, 2018.

Siena, Kevin. *Rotten Bodies: Class and Contagion in Eighteenth-Century Britain*. Yale University Press, 2019.

Sharpe, Christina. *In the Wake: On Blackness and Being*. Duke University Press, 2016.

Sheridan, Richard B. "The Guinea Surgeons on the Middle Passage: The Provision of Medical Services in the British Slave Trade." *The International journal of African historical studies* 14, no. 4 (1981): 601–625.

Steckel, Richard H., and Richard A. Jensen. "New Evidence on the Causes of Slave and Crew Mortality in the Atlantic Slave Trade." *The Journal of Economic History* 46, no. 1 (1986): 57–77.

Sussman, Charlotte. *Consuming Anxieties: Consumer Protest, Gender, and British Slavery*. Stanford University Press, 2000.

Sze, Julie. *Noxious New York: The Racial Politics of Urban Health and Environmental Justice*. MIT Press, 2006.

Thompson, Ki'Amber. "Toward a World Where We Can Breathe: Abolitionist Environmental Justice Praxis." *Annals of the American Association of Geographers* 113, no.7 (2023): 1699–1710.

Tomlins, Christopher. "Reconsidering Indentured Servitude: European Migration and the Early American Labor Force, 1600–1775." *Labor History* 42, no. 1 (2001): 5–43.

United States Environmental Protection Agency. "Cumulative Impacts Research." Last updated on February 19, 2025. https://www.epa.gov/healthresearch/cumulative-impacts-research.

Vandenbroucke, Jan Paul. "Is 'The Causes of Cancer' a Miasma Theory for the End of the Twentieth Century?" *International Journal of Epidemiology* 17, no. 4 (1988): 708–709.

Vale, Brian, and Griffith Edwards. *Physician to the Fleet: The Life and Times of Thomas Trotter, 1760–1832*. Boydell & Brewer, 2011.

Veitch, James. *A Letter to the Commissioners for Transports, and Sick and Wounded Seamen, on the Non-Contagious Nature of the Yellow Fever*. T. and G. Underwood, 1818.

Webster, Jane. "The Unredeemed Object: Displaying Abolitionist Artefacts in 2007." *Slavery & Abolition: A Journal of Slave and Post-Slave Studies* 30, no. 2 (2009): 311–325.

Wilberforce, William. *The speech of William Wilberforce, Esq., representative for the County of York, on Wednesday the 13th of May, 1789, on the question of the abolition of the slave trade. To which are added, the resolutions then moved, and a short sketch of the speeches of the other members*. Logographic Press, 1789.

Wilson, Andrew. *Rational Advice to the Military, When Exposed to the Inclemency of Hot Climates and Seasons*. W. Richardson, 1780.

Winston, Ali. "Medical Examiner Testifies Eric Garner Died of Asthma Caused by Officer's Chokehold." *The New York Times*, May 15, 2019. https://www.nytimes.com/2019/05/15/nyregion/eric-garner-death-daniel-pantaleo-chokehold.html.

Wood, Marcus. *Blind Memory*. Routledge, 2000.

Wohl, Anthony S. "The Bitter Cry of Outcast London." *International Review of Social History* 13, no. 2 (1968): 189–245.

Zimring, Carl A. *Clean and White*. New York University Press, 2016.

Zuckerman, Arnold. "Scurvy and the Ventilation of Ships in the Royal Navy: Samuel Sutton's Contribution." *Eighteenth-Century Studies* 10, no. 2 (Winter 1976–1977): 222–234.

CHAPTER 4

Anchoring the Climate Commons: Slave Ship Earth

> ...for let us make no mistake: the climate crisis is also a crisis of culture, and thus of the imagination.
> —Amitav Ghosh, *The Great Derangement*

After the devastation wrought by Hurricane Sandy in 2012, environmental journalist and activist Mike Tidwell of *The Nation* answered the question "What is a thousand-mile-wide storm pushing eleven feet of water toward our country's biggest population center saying just days before the election?" with this response: "We are all from New Orleans now."[1] In the era of climate change, Tidwell asserts a distribution of vulnerability that resonates with popular ideas about the flattening or leveling effects of our ecological conditions.[2] Here, considerations of race, class, gender, and indigeneity supposedly yield to a neo-humanism in the face of global environmental disasters. However, Tidwell's statement also obfuscates the major insights of the environmental justice movement. As environmental justice scholars like Robert D. Bullard and Beverly Wright have established, rather than creating a lateral distribution of risk "unnatural disasters" like 2005's Hurricane Katrina exacerbate the already differential dissemination of environmental hazards.[3] Because these patterns of distribution are implemented and sustained by state policy and practice, it is not coincidental that New Orleans's historically black and poor communities were among those most affected by the breaching of the

D. Leong, *Black Literature, Social Justice, and the Environment*,
https://doi.org/10.1007/978-3-032-19288-2_4

city's levees.[4] Nor is it coincidental that FEMA's response and recovery efforts within these same communities were sluggish and grossly inadequate. Tidwell's suggestion thus ignored the realities of environmental racism by capitalizing on a bankrupt race-relations paradigm. Climate change, it seems, is colorblind.

Tidwell's decision to omit any mention of race or class is all the more curious when we consider that media coverage of Hurricane Katrina was shaped largely by these very same factors. For weeks after Katrina made landfall, local, national, and even international news reports were dominated by images of black survivors trapped on rooftops, crowded into New Orleans' Superdome, or scavenging for supplies in abandoned buildings. After the flood waters receded, these were quickly replaced with images of survivors returning to decimated neighborhoods as they surveyed homes and businesses that would never be rebuilt. While the failure of local, state, and federal governments to respond to let alone prevent these catastrophes was on full display, so too were the implicit biases of the media.[5] In what would become one of the most infamous acknowledgments of both, rapper Kanye West declared during a September 2005 NBCUniversal telethon that "George Bush doesn't care about black people." His statement, one that then-President Bush would later describe as the "all-time low point" of his presidency, was preceded by his equally pointed analysis of the antiblackness structuring the government's response to and the media coverage of the Hurricane's aftermath: "I hate the way they portray us in the media…If you see a black family, it says, 'They're looting.' You see a white family, it says, 'They're looking for food.' And you know that it's been five days because most of the people are black."[6]

West's primary example referenced a pair of widely criticized photographs that despite their nearly identical compositions were circulated with two markedly different captions. The first photograph, which featured a black man wading in chest-high water with an armload of supplies, carried the following caption: "A young man walks through chest deep flood water after *looting* a grocery store."[7] The second photograph, which depicted a similar activity undertaken instead by a white man and woman, was captioned: "Two residents wade through chest-deep water after *finding* bread and soda from a local grocery store."[8] With nothing but their subjects' race to distinguish them, the photographs and their captions drew much needed attention to how blackness informs our climate change imaginary, or the "increasingly conventionalized cluster of signs, symbols, and stories" through which we understand climate change.[9] West's impassioned and unscripted comments were, in a way, a demand to take this imaginary seriously. When read against this backdrop, Tidwell's painting of climate change as a universal and universalizing force takes on new resonance.

To be sure, the "we" and "now" of Tidwell's article are only coherent when measured against the implicit "them" and "then" of New Orleans. As the proverbial canaries in the coal mine, the city's poor, black, and brown residents made legible the "slow violence" of environmental devastation.[10] Rob Nixon's insightful term captures how the sheer magnitude of climate change poses a representational challenge for defenders of the "climate commons"—or the common-pool of environmental and atmospheric resources responsible for the health of the climate. As Tidwell points out, many inhabitants of America's coastal cities are already familiar with the localized effects of climate change like drought, coastline erosion, or regional flooding. But for the climate commons to serve as a point of social cohesion (i.e., a "we"), its protection must serve the interests of even relatively well-protected and highly resourced communities. The temporal and spatial scales of climate change must be reduced to a level at which its consequences are registered as immediate, urgent, and personal (i.e., "now"). If we understand Tidwell's invocation of New Orleans as an attempt to make material these consequences, then blackness emerges as the unspoken force driving his appeal to the climate commons.[11]

Although it is *de rigueur* to begin discussions of climate change with Hurricane Katrina, I open this chapter with Tidwell's article because it illustrates the importance of attending to the imaginary of the climate commons as both a part of and distinct from climate change. It is certainly necessary to read the two together, given that interest in the climate commons developed alongside the scientific identification of anthropogenic climate change.[12] Yet to the extent that the coherence of the climate commons depends on the threat of climate change, what is actually "common" in the climate commons is too often assumed to be self-evident. The result is a rather static definition that reduces the climate commons to shared, material resources that precede or exceed the social, political, and spatial networks in which they are situated. At the same time, Tidwell's article indicates that this understanding alone is not enough to secure a commitment to the protection of the commons. What environmental activists and public policymakers refer to as a "buy-in" requires the meeting of at least three conditions: credible strategies for achieving clearly defined goals, a process for coordinating action between multiple stakeholders, and meaningful narratives attuned to cultural barriers and social norms.[13] This chapter is interested in the latter.

In this chapter, I examine how blackness mediates the imagined relationships between the "we" of the climate commons and the "now" of climate change, paying close attention to the strategies used to represent the climate commons and to solicit investment in its defense. In doing so, I read the climate commons not only as a set of shared, natural resources, but also as a symbolic formation through which an array of desires, affects, and perceptions compete to find expression. As Timothy Luke advises apropos of this formation:

> The essentially contested quality of this imaginary, and its conflicting characteristics, force one to re-examine how mediated, constructed, and rhetorical these depictions are. Such aesthetic constructs are a risky art in which climate change images can circulate as illusion, ideology, and invention as well as factual and functional scientific findings. Rather than being simple presentations of Nature as such, these efforts are often complex representations of social forces with political agendas.[14]

I begin by situating the climate commons within the longer history of commons scholarship, tracking its evolution from its roots in medieval England to its contemporary absorption in a neoliberal model known as a "distorted commons." Taking seriously historian Peter Linebaugh's proposal to treat the commons as a verb, I then explore the climate justice movement as a form of "commoning" that employs prefigurative politics to create the *communitas* implied by the notion of a commons. The aim of this section is not to adjudicate political organizing practices as they relate to climate justice, but to probe further the "we" interpellated by and through the climate commons imaginary.

Finally, to demonstrate how this interpellation is coordinated through racial blackness, I turn to two critically-acclaimed climate fiction films: Benh Zeitlin's 2012 *Beasts of the Southern Wild* and Bong Joon-ho's 2013 *Snowpiercer*. Climate fiction or "cli-fi" is especially useful for our purposes as the genre's speculative leanings tend to amplify the "conflicting characteristics" of its imaginary. Both films anchor their representations of the climate commons on a multiracial community that aspires to and draws from what Stefano Harney and Fred Moten name the "undercommons." Despite their generally inclusive and anti-capitalist ethics, the films struggle to reconcile their radical visions with the ecological legacies of the slave ship. As I show in the preceding chapters, sailors and slaves lived in common conditions of precarity, and to mark themselves as different from the enslaved, sailors systematically manipulated access to natural resources and exposure to natural elements to increase the suffering of slaves while seeking to ensure their own survival. Consequently, white racial authority became conflated with environmental control while black bodies became resources for the extraction of value. I contend that taking an "interest" in the commons—in the overlapping senses of the word as a return on investment and a sustained engagement—refers to a process of value-production that relies on black suffering. The climate crisis therefore might be conceived of as the global expansion of slave ship ecology, or a slave ship earth.

The Tragedy of the Climate Commons

Historically, the commons referred to "collective lands and resources" in medieval England that were shared by communities, each member of which had an equal stake in usage.[15] Because sustaining these pools of resources was to the advantage of all users, communities conceived collaborative social arrangements to align individual interests with those of the group. The most consistent threat to these arrangements came from the enclosure movement, which reserved formerly common spaces for private use either by community agreement, or by the mid-eighteenth century, fiat. As common lands were crucial for the subsistence of the peasant and/or working classes, the enclosure movement facilitated substantial socioeconomic transformations, including the transition from feudalism to capitalism.[16] The historical significance of the commons and enclosure was later absorbed and re-mobilized by the contemporary usage of each term, both popularized by Garrett Hardin's 1968 essay, "The Tragedy of the Commons." Hardin, an ecologist by training, conjures a metaphorical commons to connect the depletion of resources to overpopulation and the necessity of regulation. Denouncing the capacity of *Homo economicus* to manage population growth, he famously blames the welfare state for recognizing reproduction as a public right, thereby allowing large families to selfishly exploit collective resources: "Perhaps the simplest summary of this analysis of man's population problems is this: the commons, if justifiable at all, is justifiable only under conditions of low-population density."[17] To substantiate these assertions, he turns to a hypothetical group of herders and their common grazing lands. As each herder maximizes their personal profits by permitting their cows to overgraze, the commons inevitably fails.[18] Without private property rights or government intervention, he alleges, "Freedom in a commons brings ruin to all."[19]

Hardin's essay has been widely criticized by scholars from a number of disciplines, not least for its unethical and ahistorical reasoning. His embrace of neo-Malthusianism led to charges of eugencism and scientific racism, which were later borne out by his support for anti-immigration causes and race-based psychometrics.[20] Most notably, in a series of studies that would earn her the 2009 Nobel Prize in Economics political economist Elinor Ostrom demonstrated that Hardin's hypothetical commons did not accurately reflect how common-pool resources are consumed. Arguing that he "had seriously confused the concept of common property with open access conditions where no rules existed to limit entry and use," her extensive fieldwork confirmed that cooperation, collective trust, and polycentric governance are effective in preventing the "tragedies" predicted by Hardin's model.[21] Hardin himself would later revise his own understanding of the commons, observing that "the title of [his] 1968 paper should have been 'The Tragedy of the Unmanaged Commons.'"[22] Nevertheless, his error would prove productive, as his essay cemented the contemporary idea of the commons as any open-access regime containing a resource—water, air, data, health care, network access—that is or should be collectively shared. Enclosure, which was initially associated with the privatization of land, was likewise broadened to include the variable tactics of neoliberalism.

This ability to be rendered abstractly has transformed the commons into a powerful concept-metaphor, so much so that it has been appropriated by the very forces it once sought to resist.[23] Accordingly, in some currents of mainstream environmentalist discourse, "the idea of the commons seems to function less as an alternative to capitalist social relations, and more like their *savior*."[24] Many climate change solutions, for instance, prescribe some form of individual behavior modification while protecting free-market or neoliberal capitalism. The options Tidwell lists in his article to combat climate change reflect this thinking: "(1) abandon our coastal cities and retreat inland, (2) stay put and try to adapt to the menacing new conditions or (3) stop burning planet-warming fossil fuels as fast as possible."[25] Concluding that the solution lies in some combination of options 2 and 3, he suggests that we "[put] a price on carbon fuels" to reduce the costs of clean energy. The United Nations Environment Programme in fact had adopted a similar approach in 2008. The REDD (Reducing Emissions from Deforestation and Forest Degradation) and REDD + (where "plus" refers to measures that supplement conservation) programs reward countries for acting to mitigate climate change with investment incentives, some of which come in the form of carbon credits or "offsets."[26] In these "cap-and-trade" plans, governments are responsible for setting limits (i.e., "cap") on the amount of harmful emissions a corporation can produce. To comply with these limits, corporations can either: (1) find ways to reduce the amount of emissions that are generated in the course of their normal operations; (2) purchase emissions reduction credits from corporations whose overall emissions load falls below the cap; or in the case of multinational partnerships; (3) purchase carbon offsets (i.e., "trade").[27] Offsets, which involve the replacement of profits normally gained through deforestation, essentially pay developing countries not to cut down sections of their forests.

Like the Kyoto Protocol and Paris Agreement, these multilateral trade agreements conceive of the climate commons in the same economic or transactional terms that characterize Hardin's hypothetical commons. This not only justifies the use of coercive regulatory measures (e.g., investment incentives), but also discourages closer scrutiny of the local context surrounding a climate resource. As Tracy Perkins and Aaron Soto-Karin attest, "both offsets and cap-and-trade are built on the premise that as long as emissions reductions happen, it doesn't matter where they happen."[28] Their analysis of the 2010 California-Chiapas-Acre agreement—the first REDD agreement enacted between states in different nations—documents several ways that the disregard for local context leaves the underlying conditions of environmental racism intact:

> If, instead of reducing their own emissions, industries in low-income communities of color repeatedly choose to purchase credits from other facilities located in places that have more political pressure from wealthy residents, low-income communities of color could miss out on the local benefits of pollution reduction...Activists also argued that the offset system was vulnerable to abuse. It might not actually result in reductions to greenhouse gases if the state allowed offsets that did not truly reduce carbon emissions.[29]

Rather than addressing how capitalism creates and then instrumentalizes global environmental disasters, these attempts to "greenwash" the economy merely authorize the further expansion of neoliberal practice.[30] What Massimo De Angelis calls capitalism's "schizophrenic relationship to the commons" has thus produced a "distorted commons," or a commons that is paradoxically created and preserved through acts of enclosure.[31]

At the same time, the commons/enclosure dynamic also gave rise to a politics or practice of "commoning," in which a general refusal of enclosure corresponds to a preservation of the common good. Historian Peter Linebaugh, one of the most acclaimed investigators of the commons, was the first to popularize this view.[32] After locating the commons and commoning at the heart of one of Western democracy's founding documents, the Magna Carta, he concludes: "To speak of the commons as if it were a natural resource is misleading at best and dangerous at worst—the commons is an activity and, if anything, it expresses relationships in society that are inseparable from relations to nature. It might be better to keep the word as a verb, an activity, rather than as a noun, a substantive."[33] This more radical interpretation foregrounds the creative and dynamic aspects of the commons and encourages us to see it more as a social-spatial process and less as a product made available through regulation. Projects as diverse at Wikipedia (i.e., a digital knowledge commons), the Zapatista movement (i.e., a land-based, socioeconomic commons), and urban gardening (i.e., an ecological commons) have been considered examples of commoning, some in an *ex post facto* manner.[34] With respect to the climate commons, this important work has been taken up chiefly by the movements for climate justice.

To challenge the market-driven logic of the distorted commons, climate justice movements seek to apply the principles of environmental justice, democratic governance, and human rights to climate change. Unsurprisingly, indigenous communities and communities of color were the first to advocate for this approach. Their efforts had a considerable influence on the 1992 United Nations Conference on Environment and Development (UNCED), also known as the Rio Earth Summit, which became the first major international conference to acknowledge formally the detrimental effects of fossil fuels. The term "climate justice" was not widely used until the 2000 Climate Justice Summit at the Hague, which was held as an alternative to the Sixth Conference of the Parties (COP6) to the United Nations Framework Convention on Climate

Change (UNFCCC).[35] The tenets of climate justice were further elaborated by the International Climate Justice Network during the second Earth Summit in Johannesburg in 2002. In a document entitled "The Bali Principles of Climate Justice," the Network implicates "multilateral development banks, transnational corporations and Northern governments" in the "unsustainable production and consumption practice at the root" of climate change, while decrying the exclusion of "local communities, affected people and indigenous peoples" from "the global processes" intended to address it. To "build an international movement of all peoples for Climate Justice," they list 27 core principles, which include (among others): "the rights of communities dependent on natural resources for their livelihood and cultures to own and manage the same in a sustainable manner," an affirmation "that any market-based or technological solution to climate change...should be subject to principles of democratic accountability, ecological sustainability and social justice," and demands for moratoriums on "all new fossil fuel exploration and exploitation," "the construction of new nuclear power plants," and "the construction of large hydro schemes" ("Bali Principles"). These were later expanded upon via collaborative events like the 2007 founding of the Climate Justice Now! network, the 2009 Klimaforum held in parallel with the Copenhagen Climate Talks (COP15), the 2014 Flood Wall Street protests, and the Youth Climate Strikes led by Swedish activist Greta Thurnberg.

As this abridged history indicates, climate justice is capable of responding to a variety of local and translocal needs, and this flexibility stems in large part from an embrace of commoning.[36] To illustrate, in their study of COP15, Paul Chatterton et al. identify five spaces around which the talks were organized, with the third and fourth comprising a network of sites that housed the Klimaforum, climate justice activists, and various protestors.[37] Characterized by "anarchist and self-organised, autonomist politics," these sites provided "free communal sleeping spaces, kitchens, info points, indymedia centres, convergence spaces and legal information and support spaces."[38] People, ideas, and resources circulated not only within and through the sites themselves but also between organizations and communities. These exchanges would ultimately contribute to "A People's Declaration from Klimaforum09," a collaborative document conceived around the rallying cry "System change – not climate change."[39] In effect, climate justice activists instituted a spatial, social, and informational commons as a step toward the successful defense of the climate commons. Chatterton et al. explain:

> Activists accept that commoning will be contradictory and will weave together practices and values that will sometimes feel embedded or trapped in capitalist ways of doing things, and at other times will be more creative or antagonistic. This is not to say that building commons are likely, easy or free of diverse cross cutting relations of power. What is crucial is that they are prefigurative (i.e., they practice the future that they wish to see), open, experimental and have the potential to generate solidarities.[40]

Unlike the UNFCCC's approach that treats the "we" of the climate commons as external to the resources it contains, the climate justice movement seeks to build the commons through community and community through the commons. Ideally, these communities would administer their own climate change initiatives, thereby weakening the need for the coercion required by a distorted commons. Moreover, commoning can also cultivate climate change narratives that are more organic to affected communities.

In September 2005 or a little more than a month after Hurricane Katrina, nature writer Robert Macfarlane asked in *The Guardian*, "Where is the literature of climate change? Where is the creative response to what Sir David King, the government's chief scientific adviser, has famously described as 'the most severe problem faced by the world?'."[41] Although Macfarlane was concerned with the literary arts, his article was an early example of a growing interest in the climate change and climate commons imaginaries. What we needed, Macfarlane inferred, was "an imaginative repertoire…by which the causes and consequences of climate change can be debated, sensed, and communicated."[42] Since the publication of his article, the scale and scope of this repertoire have increased dramatically, leading to debates among environmental scholars and activists over its relative forms and functions. At issue are the types of data, images, stories, and symbols that are most effective at securing public support for the climate commons. Do we need better visual rhetoric for conveying scientific data?[43] Collective visions of the future?[44] Or, as per the practice of commoning, narratives that reflect the complexities of our ecological lives?

For Macfarlane, the answer lies with literature's ability to forge affective associations between the realities of climate change and their more speculative consequences. After referencing Bill McKibben's claim that "individuals would not act against climate change…until they felt 'fear in their guts,'" he proposed that "literature has a role to play in inducing this gut feeling, for one of its special abilities is that of allowing us to entertain hypothetical situations - alternative lives, or futures, or landscapes - as though they were real."[45] Despite or perhaps because of his emphasis on "hypothetical situations," Macfarlane was careful to note that climate change literature "would need to find ways of imagining which remained honest to the scientific evidence."[46] His commitment to scientific realism and politicized affect was later echoed by journalist Dan Bloom, who is widely credited with coining the term cli-fi: "For me, the best of cli-fi does two things: it delivers a powerful and emotional story and it pushes the reader to wake up to the existential threat that man-made global warming poses to future generations…If it doesn't wake us up, it's just escapist entertainment. I am not interested anymore in escapism."[47] Broadly defined as literary or cinematic works that center on anthropogenic climate change, cli-fi has become a vital site of production for the climate change and climate commons imaginaries.[48]

As evidenced by Bloom's rejection of escapism, cli-fi's appeal derives in part from its pedagogical and political potential. The genre has been lauded for its ability to "[translate] graphs and scientific jargon into experience and emotion,"[49] "[mark] the end of the idea that the nonhuman world is a God-given space for human mastery,"[50] and shorten the "'knowledge-action gap' that…prevents people from acting when they acquire frightening knowledge."[51] This is to say that cli-fi can not only concretize and personalize otherwise abstract data, but also enables us to think about the histories of the present as they relate to possible futures. My use of the plural forms of "history" and "future" is deliberate because the question of whose histories and futures are foregrounded in the climate change imaginary—and how they are represented—is precisely what is at stake. Amitav Ghosh makes a similar point in his 2016 treatise on climate change, entitled *The Great Derangement*. After identifying imperialism as central to the Anthropocene, he argues that climate change discourse is afflicted by a Eurocentrism that obscures how "the continent of Asia is conceptually critical to every aspect of global warming."[52] This Eurocentrism, he contends, stems from a reluctance toward dismantling imperialism's structures of inequality. Climate change discourse ignores "the brute fact…that no strategy can work globally unless it works in Asia" in order to pursue the neo-Malthusian politics of the distorted commons.[53]

As much as I agree with Ghosh about the Eurocentrism of climate change discourse, I cannot say the same about his assessment of cli-fi. In this chapter's epigraph, he refers to a crisis of imagination that inhibits the modern novel from dealing adequately with the Anthropocene. Because modern or "serious fiction" is fixated on the probable and pedestrian, he laments, the unpredictable consequences of climate change have been relegated to the (supposedly) lesser realms of speculative and science fiction and to cli-fi in particular.[54] What galls him most about this relegation is cli-fi's preoccupation with otherworldly futures, which he claims detracts from a more sustained reckoning with the "now" of climate change: "the Anthropocene…is precisely not an imagined 'other' world apart from ours; nor is it located in another 'time' or another 'dimension.'"[55] As literary critics have pointed out, however, these otherworldly or speculative futures are often extrapolated from the present-day experiences of poor, indigenous, and non-white communities.[56] By connecting people, places, objects, and events across time and space, cli-fi's speculative elements help us to see better the limitations and possibilities of the climate commons imaginary. In what follows, I examine how *Beasts of the Southern Wild* and *Snowpiercer* navigate these limitations and possibilities as they build a sense of belonging to a world defined by climate change.

A Light in the Wild

Beasts of the Southern Wild is set on a fictional Louisiana island named the "Isle de Charles Doucet." Better known to its residents as the "Bathtub," this bayou community is separated from the nearest urban area by a chain of levees.[57] The film's narrative unfolds from the viewpoint of a six-year-old black girl named Hushpuppy (Quvenzhané Wallis), who lives next to her ailing father, Wink (Dwight Henry), in a pair of dilapidated trailers. After a hurricane floods the area, the remaining residents are forcibly rounded up civil authorities and relocated to a refugee center. Once there the residents revolt and return to the Bathtub with a dying Wink. Hushpuppy and a group of her friends then launch themselves into the sea to find her missing mother and are delivered by boat to a brothel, where the working women lavish maternal affection on the group. Upon returning to the Bathtub, Hushpuppy confronts a herd of ancient aurochs—an extinct bovine species—that have been released from their glacial suspension by the melting of the polar ice caps.[58] Telling them, "You're my friend, kind of," she manages to avert their impending rampage in time to bid her father farewell. The film ends with a seemingly triumphant scene as Hushpuppy leads the Bathtub's residents deeper into the delta while the encroaching sea laps at their feet.

Hailed as one of the best films of its year, *Beasts* has generated critical acclaim for its treatment of climate change politics. Nicholas Mirzoeff, for example, describes *Beasts* as one of the first films to depict alternatives to neoliberal governmentality as crucial for climate change resistance. Calling the film "unbounded and undomesticated" because "the crisp, empty space of the modern cinema is here overflowing with what Jane Bennett calls 'vibrant matter,'" he suggests that contrary to colonial and modernist notions of space and time as "empty," the wild has always been "occupied."[59] As evidence he cites frames that are bursting with tight shots of "piles of crustaceans," "thickets of dense vegetation," "masses of melting glacial ice," "dust motes," and "insect life."[60] This welcome departure from the wild as either a primitive scene of violence or a space of (white) unmarked purity implies that life can thrive outside of the narrow constraints of *Homo economicus*. Nevertheless, for black feminist scholars like bell hooks, Jayna Brown, and Christina Sharpe, these riches of nature are more properly understood as the consequences of systemic racism. For hooks, the film reproduces the historical conflation of blackness with material and spiritual poverty. She insists: "there is nothing radical about the age-old politics of domination the movie espouses - insisting that only the strong survive, that disease weeds out the weak (i.e., the slaughter of Native Americans) that nature chooses excluding and including. If Wink represents the dying untamed primitive then what does Hushpuppy represent? Her fate is unclear."[61] According to hooks, the film's sharp contrast between the civility of rational society (i.e., the city protected by levees) and the Bathtub's "dying untamed primitive" endorses a survival of the fittest mentality. Climate change resistance in this sense is less a defense of the common good than it is a continuation of colonialism by other means.

These dissonant readings expose the film's struggles to reckon with the racial logic of environmental representation. In the words of Natalia Cecire, *Beasts* "aims to render slow violence representable [by] borrowing the African American strategy of producing punctual scenes that collapse and compress the temporal and spatial extension of harm."[62] To cultivate a sense of responsibility, or an environmental "non-innocence," the climate commons depends on culturally meaningful symbols to visualize the "collapsing and compressing" of ecological harm. Hence, it is not insignificant that the effects of climate change accumulate in the figure of Hushpuppy. For one, the historical refusal of innocence to black children renders them inherently culpable. The widespread criminalization of black youth is one of the more pervasive materializations of this legacy. Consequently, the racialization of innocence, or more accurately of "non-innocence," qualifies black children to bear the weight of environmental risk and responsibility.

It is not by happenstance that the film's title sequence features Hushpuppy as a literal bearer of light. Just days before a hurricane destroys most of the Bathtub, we observe Hushpuppy running toward the camera and through a darkened field with her arms outstretched, each hand holding a sparkler as the explosions from fireworks fill the frame until it brims with light (Fig. 4.1). Complimentary interpretations of this scene, and the film more broadly, claim that Hushpuppy's delight emanates from her "transcendental feelers," which guide her realization that it is "worth the risk…to live free in the riches of nature than surrounded by prosperity and plenty that's always out of reach."[63] Film critic A.O. Scott of *The New York Times* similarly argues that *Beasts* is "animated by the same spirit of freedom it sets out to celebrate."[64] While critics generally acknowledge Hushpuppy's ostensibly innocent perspective of the world, this fixation on her childhood exuberance serves a dual purpose. On the one hand, her behavior is only remarkable because viewers are unprepared to expect signs of innocence and joy in black children. On the other hand, to construe her joy as triumphant or liberating implies that it is not a joy borne of ignorance; rather, it bespeaks a precocious wisdom to which the rest of us are belated. She may be a "magical negro" in miniature, but as "caretaker, man, boy, girl, woman all within herself…part of the community but complete unto herself," she is still accountable for all the mature functions of her trope.[65]

The fascination with innocence aside, *Beasts*' portrayal of blackness participates in an economy of sentiment contingent on the possession or replacement of a suffering body. In her review of the film Jayna, Brown writes: "The film calls this poverty freedom. But I don't recognize this freedom. Their existence isn't active or sustainable…This is no maroon society…Instead the film recapitulates the continuing currency of Black suffering, and acts as a kind of 'crisis porn,' showing how black pain is erotically charged."[66] For Brown and hooks, Hushpuppy's vulnerability and need are obscured by the film's irrational celebratory spirit. The light that bursts from the frame at the end of Hushpuppy's fireworks run is not meant for her; it signifies neither the freedoms of poverty nor a rejection of governmentality. Instead, this light is meant for us, the viewers, as a distracting substitute for real political action. Fellow reviewer Christina

Fig. 4.1 Hushpuppy running through a field with sparklers

Sharpe agrees, commenting that viewing *Beasts* as enacting resistance to climate change "requires that one have no desire to alleviate Hushpuppy's devastation...The film needs black bodies because how else could incipient sexual and other violence, the violence of extreme poverty, flooding, the violence of a six-year old girl child living alone in her own ramshackle house with no mother or father, be inspiring and not tragic?."[67] In her reading, for the film to anchor the climate commons in a shared precarity it must first instrumentalize black suffering. A closer reading of the fireworks scene supports this interpretation.

Before a hurricane destroys most of the Bathtub its inhabitants gather for a large celebration. As a trawl net opens to release a writhing mass of fish, crabs, and shrimp into a wooden tub, we hear Hushpuppy say in voice-over: "One day, the storm's gonna blow, the ground's gonna sink, and the water's gonna rise up so high, there ain't gonna be no Bathtub, just a whole bunch of water."[68] The scene shifts to a darkened field, where we watch Hushpuppy point a sparkler at a grinning Wink before directing it skywards. The camera settles briefly on Hushpuppy's upturned face, lit by the glow from fireworks exploding out of frame, and over a reverse shot of Wink Hushpuppy exclaims: "But me and my daddy, we stay right here. We's who the earth is for."[69] As Hushpuppy runs through a field holding her sparklers, she points them at the camera. Light quickly fills the frame, obscuring her figure until she is hidden from view. And because the light from her sparklers continues to stretch toward the camera, the viewer can only conclude that behind this luminous opaque screen Hushpuppy continues her celebration without us. The film, in this brief but pivotal moment, refuses our attempts to identify with her. Instead, the sentiments of empathetic identification are displaced onto Hushpuppy's gift of light as the film's cinematic strategies literally wash out her racial blackness. It is only in this manner that the "we" in her earlier declaration can encompass the viewer. Because "the storm's gonna blow, the ground's gonna sink, and the water's gonna rise up," the colorblind "we" of the climate commons becomes "who the earth is for." In this register, Sharpe is correct; we cannot identify with Hushpuppy as a black girl negotiating an antiblack world, but we can identify with her when she reflects back to us the possibilities of our own survival.

While the film and its reviewers have different motivations that shape their engagements with climate change, they nonetheless converge in one implicit question: within the climate commons are black peoples interested users of an environmental resource or a resource for the generation of environmental interests? In this way, the debates over the film and the film itself not only reflect the difficulties of responding to this question, but also draw us back to the ecology of the slave ship. The correlation Sharpe alludes to between an investment in climate change resistance and a complacency toward black suffering lays bare the central challenge of the film. In visualizing the climate commons, it must avoid elevating what *is* common, namely the abjection of racial blackness, to the level of *the* commons. On this score, it is not entirely successful. Hushpuppy's affirmation that "we's who the earth is for" sounds like Tidwell's assertion that "we are all from New Orleans now." If Hushpuppy and Wink's confidence in the face of environmental precarity belongs to everyone, then the viewer is relieved from addressing the enabling conditions of that precarity. Even though the film fails to transcend the ecology of the slave ship, we can discern in this failure the possibility of a different relationship between blackness, climate change, and the climate commons imaginary.

An Undercommon Refusal

Because the distorted commons now occupies a substantial portion of the climate commons imaginary, our ability to distinguish between the two has been gradually overwhelmed. This conceptual myopia is sustained in large part by the rights-based framework prevalent in climate commons discourse. For instance, in a June 2019 press release the UN's Special Rapporteur on Human Rights and the Environment "called on countries to take urgent action to improve air quality," declaring unequivocally that "Clean Air is a Human Right."[70] Applying a human rights framework to climate resources can be a useful strategy, especially for intergovernmental organizations like the UN. But as I argue with regard to the UN's REDD programs, these rights depend on state and capitalist oversight of the commons and more specifically on the regulation of its social-spatial relations. As Stefano Harney and Fred Moten demonstrate in *The Undercommons*, a rights-based commons endorses a governing principle through which disorderly populations can be brought into the enclosures of the state. To illustrate, they turn to the frontier settlement as it is most often depicted in mainstream cinema. Appearing surrounded and besieged by indigenous peoples dispossessed of their lands, the settlement's enclosure is presented as an act of self-defense. The roles of aggressor and defender are thereby inverted and the "civil union of settlement and enclosure" convinces us not only "that we are surrounded, that we must take possession of ourselves, correct ourselves," but also that we must defend an "illusory right to what we do not have, which the settler takes for and as the commons."[71] This conflation of enclosure, defense, and possession, or what Moten elsewhere calls "the brutal ruse of self-determination," turns the self into a resource that can and must be managed.[72] However, the "we" for whom this commons is an "illusory right" is not a "we" that is easily managed. As Harney and Moten emphasize, the frontier settlement "really was surrounded, is besieged by what still surrounds it, the common beyond and beneath – before and before – enclosure."[73] *Beasts*, in this regard, is a story told from and about the surround.

In contrast to a rights-based commons, the film constructs the Bathtub as an "undercommons," or "a wild place that is not simply the left-over space that limns real and regulated zones of polite society" because it "continuously produces its own unregulated wildness."[74] Like the people, practices, and places of commoning, the undercommons is as much a *who* (e.g., black, brown, poor, indigenous, and queer peoples) and a *what* (i.e., "unregulated wildness") as it is a *where* (i.e., the "left-over" spaces of antiblackness, capital, and colonialism). Composed of those refused entrance into the commons, the undercommons refuses this refusal by working neither for the right of admission nor only for the elimination of enclosures, but for a world in which enclosures no longer make sense. In this "nonplace of abolition,"[75] Jack Halberstam explains, "we refuse to ask for recognition and instead we want to take apart, dismantle, tear down the structure that, right now, limits our ability to find each other, to see beyond it and to access the places that we know lie outside its walls."[76] What the undercommons offers therefore is an opportunity to reject the terms of a false choice: between a distorted commons or no commons at all. By projecting these terms onto the levee system and the Bathtub, *Beasts* invites viewers to consider their own investments in this choice.

The Bathtub is an allegorical reference to Louisiana's delta region and specifically to New Orleans's Ninth Ward. The Ward is located downriver of the Mississippi River watershed with large portions of it abutting the city's Industrial Canal and Gulf Intracoastal Waterways. Only the levees built and maintained by the Army Corps of Engineers protect it from storm surges and waterway flooding, making it one of the most vulnerable locations in the delta. And yet, one of *Beasts'* most striking scenes portrays the underdeveloped areas of the delta as vastly preferable to those inland. Four minutes into the film, a jump cut finds us on the waters of the Bathtub. Wink and Hushpuppy, sitting in the bed of a pickup truck detached from its cab, are floating in the waterway next to a giant set of levees. Above the levees massive industrial works are partially obscured by a haze produced by their smokestacks. Together, the industrial works and levees comprise the popular iconography of climate change—the former representing one of its causes and the latter a consequence. We shift focus from Wink, lying against the back of the cab, to Hushpuppy as she surveys the walls. Wink asks, "Ain't that ugly over there? We got the prettiest place on earth." Hushpuppy says: "Daddy says, up above the levees, on the dry side, they're afraid of the water like a bunch of babies. They built the wall that cuts us off."[77]

From the perspective of the Bathtub, the levees appear not as a life-saving adaptation for climate change, but as a tool for the ecological management of populations. The Bathtub, its inhabitants, and the water (i.e., Hushpuppy's "us") were there prior to the erection of the levees and it is fear of their "unregulated wildness" that called these concrete barricades into being. What's more, the levees impede the delta's drainage systems so that when the Bathtub floods, the excess and now contaminated water has nowhere to go. When the remaining residents survey the damage post-hurricane, we see that the once lush landscape has withered dramatically. The island's coastline is dotted with animal carcasses and the surrounding seas, the community's primary source of food, are devoid of life. When yet another fishing attempt fails, Hushpuppy asks, "You wanna leave the Bathtub, Walrus?". Walrus (Lowell Landes) responds in the negative and Wink interjects with "I heard that! Nobody's leaving the Bathtub! You don't know nothing! This water gonna kill us here. We gotta do something 'bout this. I ain't starvin' to death while them people go grocery shopping and all that."[78] In responding to Hushpuppy, Wink addresses the circumstances of her question rather than the question itself. Whatever benefits the levees provide to "them people," he stresses, tend to come at the expense of those on the outside. Wink and his friends decide to boat to the levee and blow a hole in a wall to drain the water. As Hushpuppy stows away on one of the boats, she says in voice-over: "The entire universe depends on everything fitting together just right. If you can fix the broken piece, it can all go right back."[79] At the levee, the group stuffs a fish with dynamite and attempts to latch it to the wall but lose the tripwire. Hushpuppy, who manages to retrieve it, detonates the fish and the wall explodes in a cloud of white debris. Far from contributing to the region's social and ecological well-being, the levees are the "broken piece" that must be fixed.

Having been refused entrance into this dry-land enclosure, Hushpuppy and Wink refuse this refusal by remaining in the Bathtub. This is by no means an *obvious* choice. Given the Bathtub's precarious living conditions, choosing to stay might also mean choosing poverty, illness, and high levels of environmental risk. As director Benh Zeitlin asks in *Vice*'s "The Creator's Project," why would someone "stay in a place when [they] know that the place [they] live in is doomed?".[80] At the same time, this is by no means an obvious *choice*. Relocating can be impractical or impossible for communities like the Bathtub due to any number of economic, social, and/or political factors. Using the language of choice uncritically often means ignoring the larger systems in which choices are made. What we lose as a result is a sense of the scale of those enclosures that make the distorted commons possible in the first place. More importantly, staying looks most like a choice when it is viewed from a position behind the levee walls. From this vantage point, the distorted commons and the commons can appear nearly indistinguishable. By contrast, because Hushpuppy and Wink's refusal issues from the undercommons it is less an isolated or individualized decision than it is a way of moving through and perceiving the world.

To elaborate, after the initial scene at the levee the film jumps to an aerial shot of the delta. From high above, we approach a set of makeshift dwellings arranged in a semi-circle near the edge of the water. Over a muted rumble that soon resolves into the sounds of celebration, Hushpuppy comments in voice-over: "They think we all gonna drown down here. But we ain't going nowhere."[81] We follow Hushpuppy and the other residents deeper into the Bathtub where the viewer is pulled into a raucous festival complete with a parade of floats. Like the buildings lining the street, the floats are constructed out of salvaged materials and this improvisational approach, or what Patricia Yaeger calls a "dirty ecology," is replicated in the community's social life.[82] As the celebration stretches into the night, the camera lingers on different groups of residents: a band of musicians, a multiracial group of babies crawling toward a finish line, Wink and Walrus drinking together, an older black man in overalls laughing with a group of young white men. Hushpuppy, dancing and sipping from an unmarked bottle, tells us: "Daddy always saying, that up in the dry world, they got none of what we got. They only got holidays once a year. They got fish stuck in plastic wrappers, they got their babies stuck in carriages, and chickens on sticks and all that kind of stuff."[83] Her repetition of the word "stuck" turns the dry world's benefits into limitations so that the convenient becomes wasteful (i.e., fish in plastic and chicken on sticks) and the secure becomes stifling (i.e., babies stuck in carriages). Life in the Bathtub is undeniably hard but as these scenes imply, a refusal of enclosure can make legible a wider range of connections and commonalities, including to and with the film's eponymous beasts.

Like other "allegories of the Anthropocene," *Beasts* brings together the local (i.e., the Bathtub) and the global (i.e., planet earth) by tracing the human and nonhuman entanglements through which they are connected.[84] Doing so transforms Zeitlin's earlier question into a broader inquiry about how to navigate precarity across multiple scales. Following Hushpuppy's assessment of the "dry world," the film cuts to the wooden tub of fish that precedes the fireworks scene. As Hushpuppy picks up a crab and holds it to her ear, its heartbeat swells until it is the only sound we hear. This moment recalls the film's opening scenes wherein Hushpuppy walks with a group of chickens outside of her trailer. Lifting a baby chick to her ear, she says in voice-over: "All the time, everywhere, everything's hearts is beating and squirting and talking to each other in ways we can't understand. Most of the time they probably be sayin' 'I hungry' or 'I gotta poop.' But sometimes they be talkin' in codes."[85] In each scene, the heartbeat is part of a metabolic language that situates Hushpuppy, and by extension the viewer, within a larger network of "talking" beasts. The language's signifying elements—circulation (i.e., hearts beating), energy production (i.e., hunger), and waste management (i.e., poop)—are not only indispensable for organic life, but also for the exchange of information between organisms and their environments. This is to say that if the network's language is metabolic then its conversations are about survival. These conversations are so vital that the Bathtub's resident teacher, Miss Bathsheba (Gina Montana), builds her educational philosophy around them.

After the opening credits, we join Hushpuppy and her peers in the schoolboat where Miss Bathsheba expands on these conversations for her students. A tight close-up shot of live crayfish fills the frame until the camera pulls back to reveal the students and Miss Bathsheba standing with the crayfish between them. Her lecture begins with "Meat. Meat meat meat. Every animal is made out of meat. I'm meat. Your ass is meat. Everything is part of the buffet of the universe."[86] As she informs the students that "every animal is made of meat," we see a modified shot-reverse shot of the students and a great horned owl, each gazing at the other through the bars of a cage. This same shot also places the viewer both inside and outside of the cage, underscoring how one's status as meat is preserved across shifts in position and perspective. No matter how we look at it, in other words, we are all part of the "buffet of the universe." To reinforce this point, Miss Bathsheba lifts her leg to display a tattoo of three humans with spears battling a pair of aurochs and says: "This here is an aurochs, a fierce, mean creature that walked the face of the earth back when we all lived in the caves. And they would gobble them cave-babies down right in front of the cave-baby parents. And the cavemans, they couldn't even do nothing about it because they was too poor and too small. Who here thinks the cavemen were sitting around crying like a bunch of pussies?".[87] When no one responds she lifts a fugitive crayfish from the floor and returns it to the pile. This movement effectively links crayfish to cavemen, both of whom are "too poor and too small" to avoid the dinner plate. It is with this circumstance in mind that she issues her final warning: "Y'all better think about that. Cause any day now, the fabric of the universe is coming unraveled. The ice caps gonna melt, the water's gonna rise, and everything south of the levee's going under. Ain't gonna be no Bathtub, just a whole bunch of water. Y'all better learn to survive."[88]

In these scenes, the practical and existential predicaments of climate change are coordinated through the perspective of the undercommons, which is to say through a refusal of enclosure. As I argue earlier in this section, the logic of enclosure equates survival with self-defense and self-defense with self-determination. The ability to manage and define the self is figured not only as a necessary condition of survival, but also as a standard for identifying who or what must be defended and who or what must be defended against. As a consequence, "non-Western environments, creaturely life, and peoples [are rendered] simultaneously [as] biological threats and abject figures of affectability or outer-determination rather than self-determination and interiority."[89] Unlike the "figures of affectability and outer-determination," the self-determined subject is guided by internal reason and capable of knowing and acting in its own interests. Armed with these instruments for its defense, the subject produced by the logic of enclosure is the self-enclosed, one who is autonomous and self-sufficient. In contrast, the predicament dramatized in Miss Bathsheba's aurochs fable and crayfish encounter seeks new ground for the self and survival. To formulate this predicament as a question: How does one survive when self-defense is inadequate or impossible, or when neither individual agency nor the resources of a collective are enough?

Once she connects this predicament to the melting of the ice caps, the axiom undergirding her lecture becomes clear: human and nonhuman animals are bound by a shared vulnerability that persists across time and space and between the individual and collective. Or as Zakiyyah Iman Jackson puts it, "affectability and mutual vulnerability had always already been constitutive to [our] relationality."[90]What the aurochs fable and crayfish encounter demonstrate is the futility of defending against this relationality and the vulnerability on which it is based. As an alternative, the "buffet of the universe" describes a mode of interdependence in which human and nonhuman animals become an "object of the Other's agency but in the absence of objectification [and] on a biological register."[91] Every animal in this buffet is made of "meat" and acts as a resource for the benefit of other users and as a user that benefits from the Other-as-resource. This mode of relating takes seriously the roles nonhuman animals play in our survival and sense of self and in so doing challenges the primacy of the self-determined subject. In this context, survival is predicated on a shared vulnerability rather than on a "phantasy of the will (unified and rational, self-directed subjectivity), or sovereignty, as the seat of agency."[92] We can therefore read Hushpuppy's earlier insistence that the community "ain't going nowhere" as a refusal of the logic of enclosure and the distorted versions of the self and commons that it produces. Furthermore, this refusal animates a mode of production that Harney and Moten call "governance."

As a technology of capture for immaterial labor, governance makes available for capital the social wealth of the undercommons. Because "capital cannot know directly the affect, thought, sociality, and imagination that make up the undercommon means of social reproduction, it must instead prospect for these in order to extract and abstract them as labor."[93] This prospecting, or a "kind of exploratory drilling with a responsibility bit," appears in practice as the solicitation and accumulation of interests.[94] In this respect, Harney explains, the NGO is a paragon of governance: "The NGO worker wants not to present the interests of others…but to provoke in others the self-identification of interests, to help them find their voices."[95] When individuals or communities identify the things they value, need, or feel a responsibility toward, the social life of the undercommons becomes legible as labor and as a potential source of labor power. Interests that are willingly articulated and publicly pursued are interests that can be renewed, managed, and rendered productive. Moreover, the (ostensibly) voluntary nature of this process mystifies its links to capital while endowing it with a moralizing force. Intimately related to the public or common good, governance "can be understood as the criminalisation of being without interests."[96] Consequently, a refusal of interests or responsibility appears as a threat to survival, one that unfolds through an immoral or irrational withholding of social wealth. We can see a similar process at work in the state's response to the Bathtub post-hurricane.

When the remaining residents refuse an evacuation order, the state sends agents to relocate them forcibly to the "Open Arms Processing Center for Disaster Relief and Response." Once inside the shelter, the camera pans through various scenes of trauma and displacement: refugees lying in beds, crowded into exam rooms, and lining up for food. In contrast to the warm tones and saturated colors that characterize the Bathtub, these scenes are lit and framed to evoke a quiet desperation. As she moves through harsh florescent lighting and past a series of white walls, Hushpuppy muses, "It didn't look like a prison. It looked more like a fish tank with no water. They say that we were here for our own good."[97] While she finds no visual similarities between the shelter and prison, her assessment proceeds from a prior yet unspoken comparison that hints at their underlying association. In other words, the shelter may not look like a prison but to the residents of the Bathtub it certainly feels like one. This is hardly surprising given the historical functions of the prison and hospital. These institutions of modern disciplinary power helped to create governable subjects and populations by incubating some of the earliest techniques of neoliberalism and biopolitics (e.g., surveillance, hygiene, self-management).

Although it is neither a prison nor a hospital, the shelter serves a similar function albeit through the invitation of governance. As evidenced by the reason for evacuation, meaning for the residents' "own good," governance concentrates on "the ones who manage to evade self-management in the enclosure," which is to say on those with an ability to survive in "a fish tank with no water."[98] Tellingly, when a doctor informs Wink that he needs immediate surgery he adds an admonishment to "think about what you want for your daughter."[99] After declaring, "I don't need nothing from you! You keep your hands off her! I don't need nothing from y'all," Wink is restrained by medical technicians.[100] Hushpuppy looks directly into the camera as she stands in the middle of a daycare, her stillness spliced with shots of a sedated Wink being wheeled into an adjacent room. She remarks, "When an animal gets sick here, they plug it into the wall."[101] In this setting, her reference to "the wall" recalls the earlier explosion at the levee. Now that the Bathtub's residents are within its confines, the cure for Wink's sickness of "no interests," or for his apparent refusal to take responsibility for himself or his daughter, is to plug him into the enclosure of the distorted commons—a solution that symbolically repairs the earlier breach.

By adopting the perspective of the Bathtub, *Beasts* demonstrates how a refusal of enclosure makes visible the distorted commons within the climate commons imaginary. This same perspective, it suggests, correlates with a more-than-human mode of relationality that recalibrates one's understanding of the natural world. In this sense, the film appears more interested in what this undercommon perspective reveals about the universe than it does in universalizing this perspective. To be clear, the undercommons is not uncommon; indeed, it is organic to and precedes every movement of enclosure. What is uncommon, according to the scenes in the shelter, is the preservation of this refusal and especially in the face of governance. After his surgical procedure, Wink unplugs himself from his medical monitors and collects Hushpuppy and the other children from the daycare. As they head for the exit, they are joined by the remaining residents who are executing their own escape plan. In the parking lot Wink pushes Hushpuppy onto a departing bus, shuts the door between them, and implores the driver to "put her somewhere good."[102] In the estimation of the Bathtub, the shelter is neither "somewhere good" nor working on behalf of "their own good." At the same time, because *Beasts* makes available this perspective to solicit interest in climate change, it runs the risk of augmenting if not performing the work of governance itself. As I illustrate below, this tension between the film's narrative and cinematic elements is a symptom of its ambivalence toward blackness and the ecological legacies of slavery.

The Undercommons and the Hold

At the core of the film's ambivalence is an inability to acknowledge the racial history of the undercommons. Given its choice of protagonists and geographical setting, the film clearly finds race important.[103] And yet, it fails to clarify how and why blackness informs the ethos of the Bathtub, its conditions of existence, or the refusal of enclosure that it endorses. On the level of narrative, questions pertaining to blackness are treated as if they pertain only or primarily to climate change. This anxious relation to blackness, Frank Wilderson theorizes, is part of a structural antagonism that underwrites political and cinematic discourse: "even when films narrate a story in which [blacks] are beleaguered with problems that the script insists are conceptually coherent…the nonnarrative, or cinematic, strategies of the film often disrupt this coherence by posing the irreconcilable questions of [black] political ontology – or nonontology."[104] For *Beasts*, these irreconcilable questions are precisely those that its narrative cannot address or more properly those that it must disavow. Indeed, it is because these questions are disavowed that they return symptomatically and cinematically to haunt the narrative of climate change into which blackness is subsumed.

I argue in this regard that the film's cinematic strategies transform the Louisiana Delta into a contemporary slave ship. This act of recognition, however unconscious, indicates that the slave ship drives the film's critique of the distorted commons. More specifically, what *Beasts* identifies as the practical and existential predicaments of climate change are tied to the ecology of the slave ship. In each case, a condition of shared vulnerability is amplified by ungovernable and unpredictable environmental elements. These threats to survival disrupt our concepts of self, other, and nature and in such a way that their limitations become clear. Whether aboard the slave ship or in the climate crisis, the operative question is how to redefine these concepts within the context of environmental precarity. To take our cue again from Miss Bathsheba that question might be: How does one survive when self-defense is inadequate or impossible or when neither individual agency nor the resources of a collective are enough? While this predicament confronts equally sailors and slaves and those in the undercommons and distorted commons, its consequences are unevenly distributed. As the film implies, those most affected by this asymmetry are better positioned to address it in transformative ways. The undercommons in this sense is both analogous to and derived from the hold of the slave ship. "The hold's terrible gift," Harney and Moten write, "was to gather dispossessed feelings in common, to create a new feel in the undercommons."[105] In this material and symbolic space, "the capacity to feel though others, for others to feel through you, for you to feel them feeling you, this feel of the shipped" anticipates the more-than-human mode of relationality that a refusal of enclosure enables.[106]

Instead of addressing these connections directly the film turns to magical realism to explain the provenance of the Bathtub or more accurately to explain it away. To be fair, magical realism is a valuable tool for environmental storytelling. As Ali Brox points out, "in its ability to represent slow violent environmental processes associated with anthropogenic climate change, magical realism can engage with the temporal and spatial scales necessary for depicting the state of risk and crises that our planetary present and future face."[107] Nevertheless, it is not for nothing that we are introduced to the film's magical realist elements during Miss Bathsheba's lecture on survival. As she tells the students, "Cause any day now, the fabric of the universe is coming unraveled. The ice caps gonna melt, the water's gonna rise, and everything south of the levees going under. Ain't gonna be no Bathtub, just a whole bunch of water," she points to a map of the delta showing the Bathtub beneath the chain of levees.[108] The camera zooms in on the Bathtub's location then cuts to a close-up of Hushpuppy as Miss Bathsheba says out of frame, "Y'all better learn how to survive now."[109] Hushpuppy turns her gaze to a poster of the South Pole before the film cuts to an aurochs frozen beneath a glacier. Over the faint sounds of cracking ice, she says "Way back in the day, the Aurochs was king of the world. If it wasn't for giant snowballs and the iced age, I wouldn't even be Hushpuppy. I would just be breakfast."[110]

In these scenes, the film focalizes its climate change narrative through Hushpuppy who shapes it according to her imaginative understanding of the world. To make sense of Miss Bathsheba's lesson, she situates the Bathtub within a mythical timeline that runs from "way back in the day" through the "iced aged" and to the near-future unraveling of the universe. Within this timeline, previous and current threats to survival are collapsed into the figure of the aurochs and it is Hushpuppy's gaze that appears to set them free. While this recourse to magical thinking befits someone her age, it also folds the Bathtub and its perspective into a hazy and ultimately timeless story about human resilience. This emplotment preserves the transformative potential of the undercommons without having to account for its development. And yet, the nonnarrative elements of these scenes disrupt this emplotment by bringing us back to the slave ship. As the camera cuts from the map of the delta to Hushpuppy, visible behind her is a poster featuring three human silhouettes that bear an uncanny resemblance to the slaves of the *Brooks* broadsides. Hushpuppy, standing to the right of the poster, is positioned as the fourth figure in this line of black bodies (Fig. 4.2). In passing its climate change narrative through the Bathtub and on to Hushpuppy, the film cannot help but bring the slave ship along.

This fraught relation to the slave ship is evidenced further in the film's treatment of black kinship. Without a doubt, Hushpuppy's familial bonds provide the emotional fuel for the narrative. It is noteworthy therefore that it never fully elaborates on the circumstances of Wink's illness or her mother's absence. Instead, the viewer and Hushpuppy receive euphemistic accounts of both. Wink's blood, we are told, is "eating itself" while her mother has "swum away." When we read these accounts through the framework of the slave ship, it becomes clear that this lack of explanation is also a displacement. After Miss Bathsheba's lecture Hushpuppy heads back to her trailer and calls for Wink. When he fails to return by nightfall, she traces the outline of a human figure on her mattress, lays next to it, and says, "Kids that got no Mamma, no Daddy and

Fig. 4.2 Hushpuppy standing in the schoolboat

nobody, they got to live in the woods, and eat grass, and steal underpants" (Fig. 4.3).[111] With Wink still missing the next day, she treks to and from the schoolboat on her own and then begins searching her trailer for food. The camera focuses on a memorial to her mother—a red jersey pinned to the wall with a crudely drawn outline of a face and head—as she tells us, "Everything we still got from mamma, we keep in my house. Daddy says, the first time she looked at me, it made her heartbeat so big that she thought it would blow up" (Fig. 4.4).[112] She unpins the jersey and as the camera lingers on her drawing Hushpuppy adds, "That's why she swam away." [113] On the level of narrative, these scenes depict a young girl learning to cope with the troubling absence of her parents. Afraid of becoming a kid with "no Mamma, no Daddy and nobody," Hushpuppy wrests whatever comfort she can from her simulacrum of each. On a cinematic level, however, they supply the necessary context for the film's choice of euphemisms. By reproducing her father and mother in silhouette, Hushpuppy places them alongside her in the line of black bodies that now stretches from the *Brooks* to the Bathtub. Understood in this register, Wink's affliction ("blood eating itself") and her mother's disappearance ("swimming away") take on a genealogical quality.[114]

Consistent with the film's anxious relation to blackness, Wink's illness is a form of self-cannibalization; one that weakens the links between the hold, the undercommons, and Hushpuppy by consuming the legible traces of his past. In doing so, it enables Hushpuppy to symbolize the

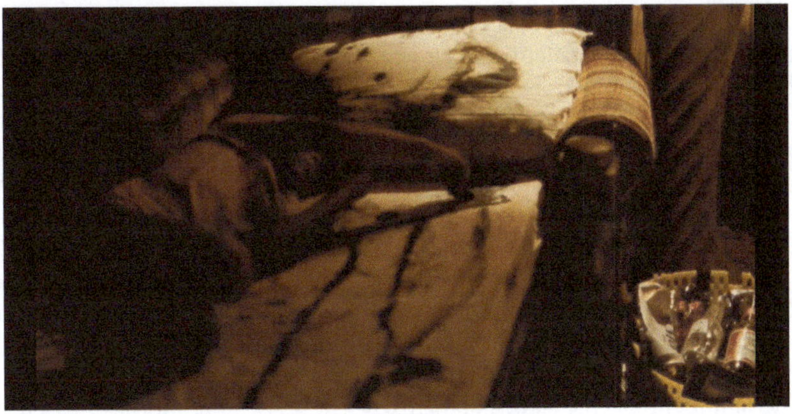

Fig. 4.3 Hushpuppy drawing an outline of Wink

Fig. 4.4 Hushpuppy's mother's red jersey

possibility of a universal climate *under*commons free from the historical burdens of race. During the film's denouement, Hushpuppy confronts the herd of aurochs that have made their way to the Bathtub from the South Pole. The lead aurochs pauses next to Wink's trailer, inspects Hushpuppy, and then bows with the rest of herd following suit. With Wink observing from his bed, she tells the aurochs, "You're my friend, kind of. I gotta take care of mine."[115] Hushpuppy enters his trailer, feeds him a bite of fried alligator, and lays her head on his chest as he dies. During these scenes, the film draws a parallel between her communion with the aurochs and her farewell to Wink. Before it turns to leave, the lead aurochs' face is captured in profile with its right eye filling the frame. This composition is replicated as Hushpuppy lays on Wink's chest with her face in profile and only her right eye in view. Each interaction may have begun with a different purpose but as these complementary shots indicate, they end in the same way—with an act of mutual recognition that neutralizes a threat.

Compelled to reference the slave ship but unable to address it, *Beasts* channels its anxieties into a campaign against blackness and its sites of reproduction. This latter objective is pursued explicitly through Hushpuppy's mother. By combining Hushpuppy's voice-over with her drawing, the film identifies the slave ship ("that's why") with her mother's absence ("she swam away"). Swimming away in this context gestures to those slaves who jumped or were thrown overboard during the crossing of the Middle Passage. Like Wink's illness, this mode of disappearance is also a defense against the racial history of the undercommons. Occurring before and after Hushpuppy's birth, her disappearance intervenes between the Bathtub and the slave ship by altering her relation to Hushpuppy. She becomes in effect both a "mother and [a] mother-dispossessed."[116] As Hortense Spillers reminds us, this paradoxical condition was established under the auspices of slavery. The legal doctrine of *partus sequitur ventrem* ("the child shall follow the womb") mandated that any child born to an enslaved woman would be classified as a slave. Accordingly, we cannot describe "'birth' in this instance as a reproduction of mothering precisely because the female, like the male, has been robbed of the parental right, the parental function."[117] With neither a legal nor a social claim to their children, black mothers and black reproduction are doubly negated.

The film reinforces this negation during the "story of [Hushpuppy's] conception," which serves as a bookend to her mother's disappearance.[118] Together these scenes effect a reversal of *partus sequitur ventrem* to ensure that Hushpuppy cannot inherit the legacies of slavery from her mother.[119] Told in a flashback, the story of her conception begins with Wink asleep in a chair and oblivious to a nearby alligator. Hushpuppy's mother steps into the frame topless and with her back to the camera, clad in the same underwear worn by Hushpuppy in the opening scenes. She cocks a shotgun and shoots the gator out of frame. With only her hips in view, she turns slowly back to the camera until it freezes on her underwear, now stained red with the gator's blood (Fig. 4.5). As the flashback ends, she is facing away from the camera and standing with her legs apart, giving us a clear view of the blood running down her legs (Fig. 4.6). While "Eurocentric environmentalism has long figured nonwhite reproductive sex as a threat to nature," *Beasts* condenses this figuration in dramatic fashion.[120] Her bloodied underwear and inner thighs are unmistakable indicators of a miscarriage or an otherwise traumatic injury to one's reproductive systems. In this way, Hushpuppy's conception and ensuing birth are marked in advance by the failure of those systems. Neither she nor her mother can become threats to "nature" or in this case to the fantasy of an undercommons disaffiliated from the hold.

Despite its celebration of the undercommons, *Beasts* can only solicit interest in its defense by reproducing the ecology of the slave ship. In essence, the film's cinematic and narrative registers assume the symbolic functions of the hold and the deck, with the former providing a steady supply of black suffering for the latter. It is by virtue of this dynamic that black suffering can simultaneously explain and conceal the film's narrative choices. As its final scenes confirm, this dual function is integral to its environmental vision. At his memorial service, Wink's body is placed into his boat and on top of a pyre. Hushpuppy sets it alight and pushes the

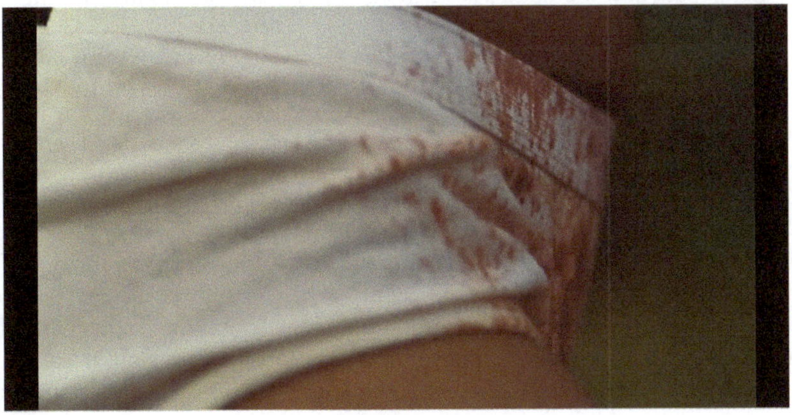

Fig. 4.5 Hushpuppy's mother with alligator blood on her underwear

Fig. 4.6 Hushpuppy's mother with alligator blood on her legs

boat away from the shore while the attendees recite a modified version of the poem "Gone From My Sight/What is Dying?":

> I am standing upon the seashore. A ship, at my side, spreads her white sails to the moving breeze and starts for the blue ocean. She is an object of beauty and strength. I stand and watch her until, at length, she hangs like a speck of white cloud just where the sea and sky come to mingle with each other. Then, someone at my side says, "There, she is gone." Gone where? Gone from my sight. That is all. She is just as large in mast, hull and spar as she was when she left my side. And, she is just as able to bear her load of living freight to her destined port. Her diminished size is in me -- not in her. And, just at the moment when someone says, "There, she is gone," there are other eyes watching her coming, and other voices ready to take up the glad shout, "Here she comes!" And that is dying.[121]

In the prevailing reading, the poem draws an analogy between the departure of a ship and the process of dying. Like the ship itself, those who leave this life may be "gone from [our] sight," but they remain "just as large" when they arrive at their "destined port." It is only from our limited perspective that their departure seems like an end rather than a transition or new beginning.

Given its central conceit, the poem is a fitting send-off for Wink. His "ship" is the same repurposed truck from which he and Hushpuppy denounce the building of the levee and from which they enact its destruction. Now carrying Wink alone, the ship's departure signals a transition between his administration of the undercommons and Hushpuppy's. Apropos of this reading, the poem is commonly misattributed to the pro-slavery politician and author, Henry Van Dyke. "Gone From My Sight," however, is the work of Luther F. Beecher, a nineteenth-century reverend, educator, and cousin to abolitionist Harriet Beecher Stowe. When approached in this light, the poem invites another interpretation. In bearing a "load of living freight," the ship at the poem's center is not only a metaphor for the passage to the afterlife, it is also a slave ship. By the same token, the poem's speaker becomes an apologist for slavery or more precisely a caricature of one, too eager to reassure us of slavery's continuation. For as the closing line declares, wherever the slave ship goes it brings death and dying with it. In short, when properly historicized, a poem reflecting on the universal experience of death is revealed as (also) a critique of the moral horrors of slavery. Taken together these two readings exemplify the film's tempestuous relation to the slave ship. By employing a poem concerned with "living freight," Wink's memorial service effectively returns him to the slave ship, and it is none other than Hushpuppy who casts both away from the Bathtub. I turn now to a brief reading of *Snowpiercer* as an alternative approach to the ecology of the slave ship.

Abolishing the Distorted Commons

Snowpiercer is set in the year 2031, after a failed attempt to combat climate change has triggered a new ice age and the extinction of almost all life on earth. What remains of the human population resides on a train that circumnavigates the globe. The train is strictly segregated by class with elites housed in luxurious cars near the engine and the poor confined to the tail sections by armed guards. Curtis (Chris Evans), the film's protagonist, instigates a revolt with the express purpose of seizing the engine, a feat that previous revolts have been unable to accomplish. He leads a group of tail passengers through the train, stopping first to collect Namgoong (Song Kang-ho) and his daughter Yona (Go Ah-sung). As the original designer of the train's security features, Namgoong assists the group in passing through several heavily fortified doors. When Curtis, Namgoong, and Yona finally reach the engine, they are met by the train's messianic engineer, Wilford (Ed Harris), who reveals that he has been collaborating with Curtis's mentor Gilliam (John Hurt) to stage periodic rebellions to reduce the train's population. He offers to let Curtis succeed him as engineer, which Curtis rejects upon learning that Wilford has been systematically replacing the train's mechanical parts with children from the tail section. Namgoong attempts to detonate an explosive substance, but the door to the engine room refuses to close. As Yona lights the fuse, Curtis and Namgoong use their bodies to shield her and Timmy, a young black boy conscripted to work in the engine, from the blast. The explosion derails the train and as Yona and Timmy emerge from the wreckage, they spot a polar bear in the distance.

Snowpiercer is the first English-language film from acclaimed South Korean director Bong Joon-Ho. Like Bong's other work—*The Host*, *Okja, Mickey-17*, and the academy award-winning *Parasite*—the film is a timely allegory about the dangers of late capitalism. Much of its critical commentary centers on its depiction of class struggle as it relates to the untenability of global petrocultures.[122] At once a symbol of modernity, the industrial revolution, and the onset of globalization, the train has been alternately described as "the contemporary neoliberal state on overdrive,"[123] a "constrained hopeless necrofuture,"[124] and a "biopolitical order [geared toward] the production and reproduction of life.[125] Already spatialized by class, the train appears to recreate in miniature the same socioeconomic dynamics that were responsible for the end of the world. Nonetheless, as Gerry Canavan points out, "This class system is retained even though class distinctions from before the disaster have no force or logic in the world afterwards. Soldiers happily enforce the feudal hierarchy even though it lacks even the slimmest relevance to the actual conditions on the train."[126] There is in fact "nothing particularly capitalistic about the political economy of the train."[127] If the train's class divisions are dissociated from its modes of production we are obliged to ask: why do they persist?

I contend that the train's configuration is modeled after the ecology of the slave ship, wherein its class distinctions conserve and conceal the value produced by its undercommon population. During the final confrontation in the engine room, Wilford observes: "Curtis, dear boy, the fact is we are all stuck inside this blasted train. We are all prisoners in this hunk of metal…this train is a closed ecosystem. We must always strive for balance. The water, the air, the food supply, the population - all must be kept in balance."[128] Like the slave ship, the train maintains its fragile ecosystem through the artificial manipulation of its resources. While the elites have access to a variety of foods, including fish, chicken, beef, and fresh vegetables from the livestock and greenhouse cars, the poor subsist on bars of gelatinized cockroaches euphemistically named "protein blocks." Along with the periodic culling of the tail sections, these regulatory practices are framed as necessary extensions of the natural. Wilford says: "We don't have time for true natural selection - we would all starve waiting for that. The next best solution is to have individual units kill off other individual units…The front and the tail are supposed to work together."[129] What he refuses to admit is that the asymmetrical distribution of resources is not an inevitable consequence of a "closed ecosystem." Rather, it is represented as such in order to naturalize the train's violent managerial strategies. And as it is with *Beasts*, the undercommons offers us the clearest perspective on these strategies.

We are first introduced to Namgoong after he has been imprisoned for his use of Chronole, a hallucinogen derived from the train's industrial waste. He agrees to join Curtis in exchange for the drug, and eventually admits that he has been collecting it not only for personal use but also for its flammable properties. Upon reaching the engine room, he points to one of the train's exit doors and says to Curtis: "We treat this like a wall because it's been frozen shut for eighteen years. But it's a door. A fucking door! Let's open it and go outside."[130] When Curtis replies, "And freeze to death? What are you, fucking crazy?", Namgoong counters, "But what if we don't freeze to death…What if we can live outside the train?".[131] The film's most radical vision comes not from Curtis, played non-coincidentally by Captain America himself, but from a non-white, former elite who refuses the logic of enclosure. His refusal is ultimately honored when Curtis sacrifices himself for the mere possibility of life outside the train. Whereas *Beasts* champions the undercommons by appropriating the ecology of the slave ship, *Snowpiercer* recognizes the "limitations of a revolution which merely takes over the existing social machinery rather than attempting to transcend it."[132]

While the train is not identical to the slave ship, *Snowpiercer* remains bound to its ecological legacies in one significant way. The engine requires a continuous supply of human "replacement parts" and the probability of acquiring them via volunteers or coercion is essentially zero. By maintaining its class divisions, the train can manufacture a steady supply of forced labor, and it is this reality that persuades Curtis to take up Namgoong's refusal of enclosure. When he and Yona join Curtis and Wilford in the engine room, Yona dislodges a floor panel to reveal Timmy wedged into the machinery. Emaciated and stupefied, he fails to respond when Yona and Curtis call to him in horror. Wilford explains, "The space only allows for a very small person. Young children under five. The engine lasts forever, but not so all of its parts. That piece of equipment when extinct recently. We needed a replacement…You know everyone has their own preordained position."[133] Enraged, Curtis punches Wilford and then sacrifices his arm to pull Timmy to safety. Notwithstanding its critique of the ecology of the slave ship, the film's final appeal for the protection of the climate commons rests on a scene of black suffering.

As works of speculative cli-fi, neither *Beasts of the Southern Wild* nor *Snowpiercer* are in the business of offering concrete solutions for climate change. Nor are they interested in providing blueprints for climate justice activism. Instead, by pursuing their environmental visons through the undercommons, the films make visible the desires, investments, and biases that circulate within and give shape to the climate commons imaginary. Doing so promotes a deeper awareness of how and why attempts to solicit interest in the environment are more or less likely to succeed. To that end, the films clarify how the ecology of the slave ship is immanent to the climate commons imaginary itself. As it is with Timmy and the train, the imaginary's inner workings are powered in part by this ecology, which connects its various components in ways that are often naturalized or disavowed. Once these connections are revealed, however, we become better attuned to the strategies and distortions of enclosure, including those that turn the hold and the undercommons into resources for the generation of environmental interests. We might therefore regard the slave ship as the first incarnation of a distorted commons, and climate justice as the pursuit of a world in which the distorted commons no longer makes sense. The aim of climate justice, in other words, is to abolish the ecology of the slave ship.

Notes

1. Mike Tidwell, "We are all from New Orleans now: Climate Change, Hurricanes, and the Fate of America's Coastal Cities," *The Nation*, October 29, 2012, https://www.thenation.com/article/archive/we-are-all-new-orleans-now-climate-change-hurricanes-and-fate-americas-coastal-cities/.
2. Christopher Caldwell, "Climate Change, The Great Leveler," *Financial Times*, December 11, 2009, https://www.ft.com/content/07feaf12-e6a8-11de-98b1-00144feab49a. In a conversation with reporters on September 12, 2005, then-President George W.

Bush commented, "The storm didn't discriminate, and neither will the recovery effort" ("Remarks on the Aftermath"). Additional comments in this vein refer to climate change and/or global warming as the "great equalizer," a "human problem," or a way to "unite humanity."

3. Manuel Pastor, Robert Bullard, James K. Boyce, Alice Fothergill, Rachel Morello-Frosch and Beverly Wright, "Environment, Disaster, and Race After Katrina" in *Race, Poverty, & the Environment* and Robert D. Bullard and Beverly Wright, *Wrong Complexion for Protection* (New York University Press, 2012).
4. Hurricane Katrina was a category 5 tropical cyclone that made landfall in Hallandale Beach, Florida, on August 25, 2005. After moving into the Gulf of Mexico and weakening into a category 3 hurricane, it again made landfall on August 29 over Louisiana and Mississippi. As the fourth-strongest Atlantic Hurricane to make landfall in the United States, Katrina was responsible for at least 1200 deaths and over $125 billion in damage, the bulk of which resulted from the catastrophic failure of New Orleans' levee system. For a comprehensive analysis of the engineering flaws that precipitated the collapse of the levees, see the American Society of Civil Engineers 2007 report, "The New Orleans Hurricane Protection System: What Went Wrong and Why," or the 2006 report issued by the National Science Foundation, "Investigation of the Performance of the New Orleans Flood Protection Systems in Hurricane Katrina on August 29, 2005."
5. Fahmy et al., "What Katrina Revealed: A Visual Analysis of the Hurricane Coverage by News Wires and US Newspapers," *Journalism & Mass Communication Quarterly* 84, no. 3 (2007): 546–561 and Voorhees et al., "'Came Hell and High Water': The Intersection of Hurricane Katrina, the News Media, Race and Poverty,'" *Journal of Community & Applied Social Psychology* 17, no. 6 (2007): 415–429. In their review of the media coverage surrounding Hurricane Katrina, Sommers et al. discovered a "disproportionate media tendency to associate Blacks with crime and violence, a propensity consistent with exaggerated and inaccurate reports regarding criminal activity in Katrina's aftermath" (1). These findings were further supported by a study of contemporaneous newspaper photographs conducted by Kahle et al., which determined that "when a person was described as looting, that

person was most often an African-American. Conversely, in all preparation photos, those showing people wisely preparing for the oncoming storm, the person in the photo was most likely to be an Anglo" (85).

6. Maxwell Strachan, "The Definitive History Of 'George Bush Doesn't Care About Black People,'" *Huffington Post*, August 28, 2015, https://www.huffpost.com/entry/kanye-west-george-bush-black-people_n_55d67c12e4b020c386de2f5e.
7. Tania Ralli, "Who's a Looter? In Storm's Aftermath, Pictures Kick Up a Different Kind of Tempest," *The New York Times*, September 5, 2005, www.nytimes.com/2005/09/05/business/whos-a-looter-in-storms-aftermath-pictures-kick-up-a-different.html (emphasis added).
8. Ralli, "Who's a Looter?" (emphasis added).
9. Timothy W. Luke, "The Climate Change Imaginary," *Current Sociology* 63, no.2 (2015): 280–296.
10. Rob Nixon, *Slow Violence and the Environmentalism of the Poor* (Harvard University Press, 2011), 3. Rob Nixon defines "slow violence" as "a violence that occurs gradually and out of sight, a violence of delayed destruction that is dispersed across time and space, an attritional violence that is typically not viewed as violence at all."
11. To be clear, I am arguing neither that Tidwell endorses a color-blind model of climate change, nor that his instrumentalization of blackness and black suffering is deliberate. Indeed, as a long-time environmental journalist and activist, Tidwell has a robust history of advocating for environmental justice. My reading of his article is intended instead to uncover how blackness supplies a grammar for determining what is "common" in the climate commons. For an overview of Tidwell's work on environmental justice, see https://chesapeakeclimate.org/mission/.
12. The earliest published use of the term "atmospheric commons" is likely from a 1973 textbook entitled *Environmental Administration* by Stahrl Edmunds, while the term "climate commons" appears first in a 1990 Staff Paper Series from the Department of Agricultural and Applied Economics at the University of Minnesota (See Ford C. Runge, "Common Property Resources in a Global Context"). A Google ngram search for the term "climate commons" also reveals a dramatic net increase in rate of

use between the years 1990 and 2019, with the year 2019 registering a ngram value of 0.0000000458 across all books published in English.

13. Simon Bushella, Géraldine Satre Buisson, Mark Workman, Thomas Colley, "Strategic Narratives in Climate Change: Towards a Unifying Narrative to Address the Action Gap on Climate Change," *Energy Research & Social Science* 28 (2017): 39–49.
14. Luke, "Climate Change Imaginary," 3.
15. Jeff Shantz, *Commonist Tendencies: Mutual Aid Beyond Communism* (Punctum Books, 2013), 3.
16. Marx outlines the contributions of the enclosure movement to the development of primitive accumulation and the labor and commodity markets in "Expropriation of the Agricultural Population from the Land," chapter 27 of *Capital*, vol. 1.
17. Garrett Hardin, "The Tragedy of the Commons," *Science* 162, no. 3859 (1968): 1248. *Homo economicus*, meaning "economic man," refers to a concept in which man is defined as rational, self-interested, and guided chiefly by the desire to maximize and/or optimize his own material conditions.
18. Rob Nixon, "Neoliberalism, Genre, and 'The Tragedy of the Commons,'" *PMLA* 127, no. 3 (2012): 594. The parable of the cow herders was borrowed from William Forster Lloyd's 1832 lecture to Oxford University, entitled "Two Lectures on the Checks to Population." This lecture, which was published as a pamphlet in 1833, influenced Hardin's description of human greed as the moral downfall of the commons.
19. Hardin, "The Tragedy of the Commons," 1244.
20. The Southern Poverty Law Center describes Hardin as a lifelong white nationalist who endorsed a range of eugenicist, nativist, and white supremacist positions. These views were made apparent in his 1974 essay, "Lifeboat Boat Ethics: A Case Against Helping the Poor," which extended the critique of the welfare state began in "The Tragedy of the Commons." See his profile at: https://www.splcenter.org/fighting-hate/extremist-files/individual/garrett-hardin
21. Elinor Ostrom, Thomas Dietz, Nives Dolšak, Paul C. Stern, Susan Ed Stonich, and Elke U. Weber, *The Drama of the Commons* (National Academy Press, 2002), 11.

22. Garrett Hardin, "The Tragedy of the Unmanaged Commons: Population and Disguises of Providence," in *Commons Without Tragedy*, ed. Dalamin Andelson, R.V. (Shepheard-Walwyn, 1991), 178.
23. Timothy W. Luke, "On Environmentality: Geo-power and Eco-knowledge in the Discourses of Contemporary Environmentalism," *Cultural Critique* 31 (1995): 57–81; Arun Agrawal, *Environmentality: Technologies of Government and the Making of* Subjects (Duke University Press, 2005); and Robert Fletcher and Jose A. Cortes-Vazquez, "Beyond the Green Panopticon: New Directions in Research Exploring Environmental Governmentality," *Environment and Planning E: Nature and Space* 3, no. 2 (2020): 289–299. As it is with other areas of neoliberal intervention, the climate commons provides an opportunity to align individual interests with those of various political and economic actors. This alignment is achieved frequently through what Timothy Luke and Arun Agrawal call environmentality, whereby the environment becomes the conceptual framework for the operations of biopower and governmentality. The construction of environmental subjects occurs in part via the creation of environmental knowledges, which are themselves integrated into the disciplinary practices of the state.
24. Massimo De Angelis, "The Tragedy of the Capitalist Commons," *Turbulence: Ideas for Movement* 5 (December 2009): 32.
25. Tidwell, "From New Orleans Now."
26. For an in-depth look at the REDD and REDD + programs, see https://www.un-redd.org/
27. Tracy Perkins and Aaron Soto-Karlin, "Situating Global Policies within Local Realities." In *Sustainability: Approaches to Environmental Justice and Social Power* (Duke University Press, 2019).
28. Perkins and Soto-Karlin, "Situating," 106.
29. Perkins and Soto-Karlin, "Situating," 107–108.
30. Naomi Klein, *The Shock Doctrine: The Rise of Disaster Capitalism* (Picador, 2007). Naomi Klein's *The Shock Doctrine* is the first book-length study on how neoliberal, free-market practices capitalize on large-scale disasters to advance exploitative economic policies.
31. De Angelis, "Tragedy of the Capitalist Commons," 33.

32. In addition to Linebaugh and Ostrom, political philosophers Michael Hardt and Antonio Negri are also considered preeminent theorists of the commons and commoning. In fact, Hardt and Negri's volume on the commons, *Commonwealth*, was published only a year after Linebaugh's *The Magna Carta Manifesto.*
33. Peter Linebaugh, *The Magna Carta Manifesto: Liberties and Commons for All* (University of California Press, 2008), 279.
34. Jeff Shantz, *Commonist Tendencies: Mutual Aid Beyond Communism* (Punctum Books, 2013); Peter Linebaugh, *Stop, Thief!: The Commons, Enclosures, and Resistance* (PM Press, 2014).
35. The UNFCCC is an international environmental treaty focused on the reduction of greenhouse gases and the mitigation of climate change. Developed and adopted at the 1992 Rio Earth Summit, the treaty and its subsequent conferences are the primary organs for international cooperation around climate issues. Both the 1997 Kyoto Protocol and the 2015 Paris Agreement are extensions of the original treaty, with the Kyoto Protocol representing the first legally binding agreement on the reduction of greenhouse gas emissions. Because representatives from the Convention's participating countries meet annually, the COP gatherings have become opportunities for climate justice activists to push for more radical solutions.
36. Shelley Streeby, *Imagining the Future of Climate Change*, Vol. 5, University of California Press, 2018; David Schlosberg and Lisette B. Collin, "From Environmental to Climate Justice: Climate Change and the Discourse of Environmental Justice," *Wiley Interdisciplinary Reviews: Climate Change* 5, no.3 (2014): 359–374; Mary Robinson, *Climate Justice: Hope, Resilience, and the Fight for a Sustainable Future* (Bloomsbury Publishing USA, 2018).
37. Paul Chatterton et al., "Articulating Climate Justice in Copenhagen: Antagonism, the Commons, and Solidarity," *Antipode* 45, no. 3 (2013): 2–4. The other three sites include: (1) an official space composed of national representatives and pre-approved NGOs; (2) a corporate/business space lobbying for market-based and techno-scientific solutions to climate change; and (3) the Copenhagen streets themselves, which hosted a range of sanctioned and non-sanctioned events.
38. Chatterton et al., "Articulating Climate Justice," 3.

39. For the full text of "A People's Declaration from Klimaforum09" see: http://klimaforum.org/declaration_english.pdf.
40. Chatterton et al., "Articulating Climate Justice," 10.
41. Robert Macfarlane, "The Burning Question," *The Guardian*, September 23, 2005. www.theguardian.com/books/2005/sep/24/featuresreviews.guardianreview29.
42. Macfarlane, "Burning Question."
43. See Luke's "The Climate Change Imaginary."
44. See Manjana Milkoreit, "Imaginary Politics: Climate Change and Making the Future," *Elementa: Science of the Anthropocene*, 5 (2017): 62.
45. Macfarlane, "Burning Question."
46. Macfarlane, "Burning Question."
47. Dan Bloom, "Q&A with Dan Bloom, Popularizer of Cli- Fi." Interview by Robin Bates. *Teleread*, January 27, 2016, https://teleread.com/qa-dan-bloom-popularizer-cli-fi-climate-fiction/index.html.
48. Adeline Johns-Putra, "Climate Change in Literature and Literary Studies: From Cli-fi, Climate Change Theater and Ecopoetry to Ecocriticism and Climate Change Criticism," *Wiley Interdisciplinary Reviews: Climate Change* 7, no. 2 (2016): 266–282. Adeline Johns-Putra's (2016) review of cli-fi provides a helpful introduction to the genre and to the broader traditions of climate change representation in environmental literature and literary studies.
49. Rebecca Tuhus-Dubrow, "Cli- Fi: Birth of a Genre," *Dissent*, Summer 2013. www.dissentmagazine.org/article/cli-fi-birth-of-a-genre.
50. Gregers Andersen, "Cli-fi and the Uncanny," *ISLE: Interdisciplinary Studies in Literature and Environment* 23, no. 4 (2016): 860.
51. Debra J. Rosenthal, "Climate-Change Fiction and Poverty Studies: Kingsolver's *Flight Behavior*, Diaz's *Monstro*, and Bacigalupi's *The Tamarisk Hunter*," *ISLE: Interdisciplinary Studies in Literature and Environment* 27, no. 2 (2020), 4.
52. Amitav Ghosh, *The Great Derangement: Climate Change and the Unthinkable* (Penguin, 2018), 87.
53. Ghosh, *Great Derangement*, 90.
54. Ghosh, *Great Derangement*, 24.

55. Ghosh, *Great Derangement*, 72–73.
56. Shelley Streeby, *Imagining the Future of Climate Change: World-making through Science Fiction and Activism*. Vol. 5. Univ of California Press, 2018; Rebecca Evans, "Fantastic Futures? Cli-fi, Climate Justice, and Queer Futurity," *Resilience: A Journal of the Environmental Humanities* 4, no. 2–3 (2017): 94–110. Shelly Streeby's study of climate change in science fiction tracks the multiple ways that "people of color and Indigenous people use science fiction and other speculative genres to remember the past and imagine futures that help us think critically about the present and connect climate change to social movements" (5).
57. The film is an adaptation of Lucy Alibar's one-act play *Juicy and Delicious*.
58. Aurochs are the extinct, wild-type ancestors of modern cattle.
59. Nicholas Mirzoeff, "Becoming Wild," *Occupy 2012* (2012).
60. Mirzoeff, "Becoming Wild."
61. bell hooks, "No Love in the Wild." Ne*w Black Man (in Exile)*. Blogspot, September 5, 2012.
62. Natalie Cecire, "Environmental Innocence and Slow Violence," *WSQ: Women's Studies Quarterly* 43, no. 1–2 (Spring/Summer 2015): 172.
63. David Edelstein, "Movie Review: Weighing the Ambitious Beasts of the Southern Wild," *Vulture*, June 28, 2012. https://www.vulture.com/2012/06/movie-review-beasts-of-the-southern-wild.html, and Keith Phipps, "Beasts of the Southern Wild," *A.V. Club*, June 28, 2012, https://www.avclub.com/beasts-of-the-southern-wild-1798173307.
64. A.O. Scott, "She's the Man of This Swamp," *The New York Times*. 27 June 2012: C1.
65. Christina Sharpe, "Beasts of the Southern Wild—The Romance of Precarity I," *Social Text* 27 (2013). For more on the "magic negro" trope in film and literature, see Marvin D. Jones, *Race, Sex, and Suspicion: The Myth of the Black Male* (Westport, CT: Greenwood Publishing Group, 2005).
66. Jayna Brown, "Beasts of the Southern Wild–The Romance of Precarity II," *Social Text* 9 (2013).
67. Sharpe, "Romance of Precarity I."
68. *Beasts*, 0:06:34–0:06:47.
69. *Beasts*, 0:07:32–0:07:37.

70. "Clear Air is a Human Right," *United Nations Climate Change*, June 4, 2019. https://unfccc.int/news/clean-air-is-a-human-right-un-special-rapporteur.
71. Stefano Harney and Fred Moten, *The Undercommons: Fugitive Planning and Black Study* (NY: Minor Compositions, 2013), 18.
72. Fred Moten, *Stolen Life: Consent Not to be a Single Being* (Durham, NC: Duke University Press, 2018), 212.
73. Harney and Moten, *Undercommons*, 17.
74. Jack Halberstam, Introduction to *The Undercommons: Fugitive Planning and Black Study* by Stefano Harney and Fred Moten, 7. Durham, NC: Duke University Press, 2018.
75. Harney and Moten, *Undercommons*, 42.
76. Halberstam, "Introduction," 6.
77. *Beasts*, 0:04:40–0:04:54.
78. *Beasts*, 0:45:47–0:46:20.
79. *Beasts*, 0:49:04–0:49:12.
80. The Creator's Project, "Making of Beasts of the Southern Wild—Part I," YouTube video, 10:29, August 15, 2012, https://www.youtube.com/watch?v=7OUM1Yy6glM.
81. *Beasts*, 0:05:00–0:05:05.
82. Patricia Yaeger, "*Beasts of the Southern Wild* and Dirty Ecology." *Southern Spaces* 13 (2013): 155.
83. *Beasts*, 0:05:45–0:06:21.
84. Elizabeth DeLoughrey, *Allegories of the Anthropocene* (Durham: Duke University Press, 2019).
85. *Beasts*, 0:02:26–0:02:45.
86. *Beasts*, 0:08:22–0:08:40.
87. *Beasts*, 0:08:44–0:09:13.
88. *Beasts*, 0:09:18–0:09:33.
89. Zakiyyah Iman Jackon, *Becoming Human: Matter and Meaning in an Antiblack World* (New York University Press, 2020), 131.
90. Jackson, *Becoming Human*, 146.
91. Jackson, *Becoming Human*, 144.
92. Jackon, *Becoming Human*, 119.
93. Harney and Moten, *Undercommons*, 80.
94. Harney and Moten, *Undercommons*, 54.
95. Stefano Harney, "Abolition and the General Intellect," *Generation*, http://dev.autonomedia.org/node/10953.
96. Harney and Moten, *Undercommons*, 57.

97. *Beasts*, 1:02:15–1:02:27.
98. Harney and Moten, *Undercommons*, 52.
99. *Beasts*, 1:03:17–1:03:20.
100. *Beasts*, 1:03:23–1:03:40.
101. *Beasts*, 1:05:04–1:05:13.
102. *Beasts*, 1:07:39–1:07:41.
103. The Bathtub is modeled on a real-world location and community in southern Louisiana called the Isle de Jean Charles.
104. Frank Wilderson, *Red, White, & Black* (Durham: Duke University Press, 2011), 5.
105. Harney and Moten, *Undercommons*, 97.
106. Harney and Moten, *Undercommons*, 98.
107. Ali Brox, "The Monster of Representation: Climate Change and Magical Realism in '*Beasts of the Southern Wild*,'" *The Journal of the Midwest Modern Language Association* 49, no. 1 (2016): 141.
108. *Beasts*, 0:09:18–0:09:33.
109. *Beasts*, 0:09:30–0:09:33.
110. *Beasts*, 0:09:42–0:09:58.
111. *Beasts*, 0:11:12–0:11:27.
112. *Beasts*, 0:12:51–0:13:06.
113. *Beasts*, 0:13:07–0:13:09.
114. *Beasts*, 1:08:14–1:08:17.
115. *Beasts*, 1:22:10–1:22:25.
116. Hortense Spillers, "Mama's Baby, Papa's Maybe: An American Grammar Book," *Diacritics* 17, no. 2 (1987): 80.
117. Spillers, "Mama's Baby," 77–78.
118. *Beasts*, 0:35:29–0:35:34.
119. Hushpuppy's mother is essentially anatomized across the stories of her disappearance and Hushpuppy's conception. In the former she appears only as a face and head while in the latter she is shot entirely from the neck down. This precise distribution suggests these scenes should be read together for a fuller understanding of her role in the film.
120. Tavia Nyong'o, "Little Monsters: Race, Sovereignty, and Queer Inhumanisms in *Beasts of the Southern Wild*," Queer Inhumanisms. Spec. issue of *GLQ: A Journal of Lesbian and Gay Studies* 21, no. 2/3 (2015): 261.
121. Luther F. Beecher, "What is Dying?", *Northwestern Christian Advocate*, July 13, 1904.

122. Stephanie LeMenager, *Living Oil: Petroleum Culture in the American Century* (Oxford University Press, 2013).
123. Rob Wilson, "*Snowpiercer* as Anthropoetics: Killer Capitalism, the Anthropocene, Korean-Global Film," *boundary 2* 46, no. 2 (2019): 203.
124. Gerry Canavan, "'If the Engine Ever Stops, We'd All Die': *Snowpiercer* and Necrofuturism," *Paradoxa* 24 (2014): 22.
125. Fred Lee and Steven Manicastri, "Not All are Aboard: Decolonizing Exodus in Joon-ho Bong's *Snowpiercer*," *New Political Science* 40, no. 2 (2018): 215.
126. Canavan, "Engine," 15.
127. Canavan, "Engine," 16.
128. *Snowpiercer*, 1:38:40–1:39:00.
129. *Snowpiercer*, 1:39:16–1:40:13.
130. *Snowpiercer*, 1:34:08–1:34:15.
131. *Snowpiercer*, 1:34:16–1:34:25.
132. Peter Frase, "Smash the Engine," *Jacobin*, July 3, 2014, www.jacobinmag.com/2014/07/smash-the-engine/.
133. *Snowpiercer*, 1:51:09–1:53:45.

References

Agrawal, Arun. *Environmentality: Technologies of Government and the Making of Subjects*. Duke University Press, 2005.

Andersen, Gregers. "Cli-fi and the Uncanny." *ISLE: Interdisciplinary Studies in Literature and Environment* 23, no. 4 (2016): 855–866.

Beecher, Luther F. "What is Dying?." *Northwestern Christian Advocate*, July 13, 1904.

Bloom, Dan. "Q&A with Dan Bloom, Popularizer of Cli- Fi." Interview by Robin Bates. *Teleread*, January 27, 2016. https://teleread.com/qa-dan-bloom-popularizer-cli-fi-climate-fiction/index.html.

Brown, Jayna. "Beasts of the Southern Wild–The Romance of Precarity II." *Social Text* 9 (2013).

Brox, Ali. "The Monster of Representation: Climate Change and Magical Realism in '*Beasts of the Southern Wild*.'" *The Journal of the Midwest Modern Language Association* 49, no. 1 (2016): 139–155.

Bullard, Robert Doyle, and Beverly Wright. *The Wrong Complexion for Protection: How the Government Response to Disaster Endangers African American Communities*. New York University Press, 2012.

Bush, George W. "Remarks on the Aftermath of Hurricane Katrina and an Exchange With Reporters in New Orleans, Louisiana." The American Presidency Project. UC Santa Barabara. https://www.presidency.ucsb.edu/documents/remarks-the-aftermath-hurricane-katrina-and-exchange-with-reporters-new-orleans-louisiana.

Bushell, Simon, et al. "Strategic Narratives in Climate Change: Towards a Unifying Narrative to Address the Action Gap on Climate Change." *Energy Research & Social Science* 28 (2017): 39–49.

Caldwell, Christopher. "Climate Change, The Great Leveler." *Financial Times*, December 11, 2009. https://www.ft.com/content/07feaf12-e6a8-11de-98b1-00144feab49a.

Canavan, Gerry. "If the Engine Ever Stops, We'd All Die": Snowpiercer and Necrofuturism." *Paradoxa* (2014).

Cecire, Natalie. "Environmental Innocence and Slow Violence." *WSQ: Women's Studies Quarterly* 43, no. 1–2 (Spring/Summer 2015): 172.

Chatterton, Paul, David Featherstone, and Paul Routledge. "Articulating Climate Justice in Copenhagen: Antagonism, the Commons, and Solidarity." *Antipode* 45, no. 3 (2013): 602–620.

De Angelis, Massimo. "The Tragedy of the Capitalist Commons," *Turbulence: Ideas for Movement* 5 (December 2009).

DeLoughrey, Elizabeth M. *Allegories of the Anthropocene*. Duke University Press, 2019.

Edelstein, David. "Movie Review: Weighing the Ambitious Beasts of the Southern Wild." *Vulture*, June 28, 2012. https://www.vulture.com/2012/06/movie-review-beasts-of-the-southern-wild.html.

Evans, Rebecca. "Fantastic Futures? Cli-fi, Climate Justice, and Queer Futurity." *Resilience: A Journal of the Environmental Humanities* 4, no. 2–3 (2017): 94–110.

Fahmy, Shahira, James D. Kelly, and Yung Soo Kim. "What Katrina Revealed: A Visual Analysis of the Hurricane Coverage by News Wires and US Newspapers." *Journalism & Mass Communication Quarterly* 84, no. 3 (2007): 546–561.

Fletcher, Robert and Jose A. Cortes-Vazquez, "Beyond the Green Panopticon: New Directions in Research Exploring Environmental Governmentality." *Environment and Planning E: Nature and Space* 3, no. 2 (2020): 289–299.

Frase, Peter. "Smash the Engine." *Jacobin*, July 3, 2014. http://www.jacobinmag.com/2014/07/smash-the-engine/.

Ghosh, Amitav. *The Great Derangement: Climate Change and the Unthinkable*. Penguin UK, 2018.

Halberstam, Jack. Introduction to *The Undercommons: Fugitive Planning and Black Study* by Stefano Harney and Fred Moten, 5–12. Duke University Press, 2018.

Hardin, Garrett. "The Tragedy of the Unmanaged Commons: Population and Disguises of Providence." In *Commons Without Tragedy,* edited by Dalamin Andelson. Shepheard-Walwyn, 1991.

Hardt, Michael and Antonio Negri. *Commonwealth.* Harvard University Press, 2009.

Harney, Stefano. "Abolition and the General Intellect." *Generation.* https://www.generation-online.org/c/fc_rent13.html.

Harney, Stefano and Fred Moten. *The Undercommons: Fugitive Planning & Black Study.* Minor Compositions, 2013.

hooks, bell. "No Love in the Wild." *New Black Man (in Exile)*, September 5, 2012.

Jackson, Zakiyyah Iman. *Becoming Human: Matter and Meaning in an Antiblack World.* New York University Press, 2020.

Johns-Putra, Adeline. "Climate Change in Literature and Literary studies: From Cli-fi, Climate Change Theater and Ecopoetry to Ecocriticism and Climate Change Criticism." *Wiley Interdisciplinary Reviews: Climate Change* 7, no. 2 (2016): 266–282.

Jones, D. Marvin. *Race, Sex, and Suspicion: The Myth of the Black Male.* Greenwood Publishing Group, 2005.

Klein, Naomi. *The Shock Doctrine: The Rise of Disaster Capitalism.* Picador, 2007.

Lee, Fred and Steven Manicastri. "Not All are Aboard: Decolonizing Exodus in Joon-ho Bong's *Snowpiercer.*" *New Political Science* 40, no. 2 (2018): 211–226.

LeMenager, Stephanie. *Living Oil: Petroleum Culture in the American Century.* Oxford University Press, 2013.

Linebaugh, Peter. *The Magna Carta Manifesto: Liberties and Commons for All.* University of California Press, 2008.

Linebaugh, Peter. *Stop, Thief!: The Commons, Enclosures, and Resistance.* PM Press, 2014.

Luke, Timothy W. "The Climate Change Imaginary." *Current Sociology* 63, no. 2 (2015): 280–296.

Luke, Timothy W. "On Environmentality: Geo-power and Eco-knowledge in the Discourses of Contemporary Environmentalism." *Cultural Critique* 31 (1995): 57–81.

Macfarlane, Robert. "The Burning Question." *The Guardian*, September 23, 2005, www.theguardian.com/books/2005/sep/24/featuresreviews.guardianreview29.

Milkoreit, Manjana. "Imaginary Politics: Climate Change and Making the Future." *Elementa: Science of the Anthropocene* 5 (2017): 62.

Moten, Fred. *Stolen Life: Consent Not to be a Single Being.* Duke University Press, 2018.

Nixon, Rob. "Neoliberalism, Genre, and 'The Tragedy of the Commons.'" *PMLA* 127, no. 3 (2012): 593–599.

Nixon, Robert. *Slow Violence and the Environmentalism of the Poor*. Harvard University Press, 2011.

Nyong'o, Tavia. "Little Monsters: Race, Sovereignty, and Queer Inhumanisms in *Beasts of the Southern Wild*." *GLQ: A Journal of Lesbian and Gay Studies* 21, no. 2–3 (2015): 249–272.

Ostrom, Elinor. *Governing the Commons: The Evolution of Institutions for Collective Action*. Cambridge University Press, 1990.

Ostrom, Elinor et al. *The Drama of the Commons*. National Academy Press, 2002.

Pastor, Manuel, et al. "Environment, Disaster, and Race after Katrina." *Race, Poverty & the Environment* 13, no.1 (2006): 21–26.

Perkins, Tracy, and Aaron Soto-Karlin. "Situating Global Policies within Local Realities." In *Sustainability: Approaches to Environmental Justice and Social Power*. Duke University Press, 2019.

Phipps, Keith. "Beasts of the Southern Wild." *A.V. Club*, June 28, 2012, https://www.avclub.com/beasts-of-the-southern-wild-1798173307.

Rauli, Tania. "Who's a Looter? In Storm's Aftermath, Pictures Kick Up a Different Kind of Tempest." *The New York Times*. September 5, 2005. www.nytimes.com/2005/09/05/business/whos-a-looter-in-storms-aftermath-pictures-kick-up-a-different.html.

Robinson, Mary. *Climate Justice: Hope, Resilience, and the Fight for a Sustainable Future*. Bloomsbury Publishing USA, 2018.

Rosenthal, Debra J. "Climate-Change Fiction and Poverty Studies: Kingsolver's '*Flight Behavior*,' Diaz's '*Monstro*', and Bacigalupi's 'The Tamarisk Hunter.'" *ISLE: Interdisciplinary Studies in Literature and Environment* 27, no.2 (2020): 268–286.

Schlosberg, David, and Lisette B. Collins. "From Environmental to Climate Justice: Climate Change and the Discourse of Environmental Justice." *Wiley Interdisciplinary Reviews: Climate Change* 5, no.3 (2014): 359–374.

Shantz, Jeff. *Commonist Tendencies: Mutual Aid Beyond Communism*. Punctum Books, 2013.

Sharpe, Christina. "Beasts of the Southern Wild—The Romance of Precarity I." *Social Text* 27 (2013).

Spillers, Hortense. "Mama's Baby, Papa's Maybe: An American Grammar Book." *Diacritics* 17, no. 2 (1987): 65–81.

Strachan, Maxwell. "The Definitive History Of 'George Bush Doesn't Care About Black People.'" *Huffington Post*, August 28, 2015. https://www.huffpost.com/entry/kanye-west-george-bush-black-people_n_55d67c12e4b020c386de2f5e.

Streeby, Shelley. *Imagining the Future of Climate Change: World-Making through Science Fiction and Activism*. Vol. 5. University of California Press, 2018.

Tidwell, Mike. "We are all from New Orleans now: Climate Change, Hurricanes, and the Fate of America's Coastal Cities." *The Nation*, October 29, 2012. https://www.thenation.com/article/archive/we-are-all-new-orleans-now-climate-change-hurricanes-and-fate-americas-coastal-cities/.

Tuhus-Dubrow, Rebecca. "Cli- Fi: Birth of a Genre." *Dissent*, Summer 2013. www.dissentmagazine.org/article/cli-fi-birth-of-a-genre.

Voorhees, Courte CW, John Vick, and Douglas D. Perkins. "'Came Hell and High Water': The Intersection of Hurricane Katrina, the News Media, Race and Poverty." *Journal of Community & Applied Social Psychology* 17, no. 6 (2007): 415–429.

Wilson, Rob. "*Snowpiercer* as Anthropoetics: Killer Capitalism, the Anthropocene, Korean-Global Film." *boundary 2* 46, no. 3 (2019): 199–218.

Yaeger, Patricia. "*Beasts of the Southern Wild* and Dirty Ecology." *Southern Spaces* 13 (2013): 155.

CHAPTER 5

The Mattering of Black Lives: Octavia Butler's *Parable* Duology

> ...transformed *meanings* have led to transformed *matter*, to a transformed mode of experiencing the *self*.
> — Sylvia Wynter, "Towards the Sociogenic Principle"

Black lives matter and black *lives* matter and black lives *matter*. This homographic reading of the most salient political statement of recent years speaks to the torsions of blackness, matter, and life that have come to define our contemporary era. In "Unbearable Blackness," Jared Sexton argues with regard to the triangulation of these concerns that antiblack fantasies "do not render blacks, like so much of the planet, subject to death in an economy of disposability; rather, they subject blacks to 'the interminable time of meaningless, impersonal dying.'"[1] Given the ongoing failure to prosecute fully the murders of black people, Sexton's claims register most strongly in the state's refusal to allow these deaths to die. They are caught instead in a biopolitical apparatus that suspends racial blackness between a life unrecognized as such and an illegible form of death that can never pass into reason. Against this timeless, spectral dying, we can read the declaration that Black Lives Matter as a call to return racial blackness to a form that matters or to a form that *is* matter. On this score, I ask: how do black life and death become matter, and what is at stake in the demand that they should assume such form?

D. Leong, *Black Literature, Social Justice, and the Environment*,
https://doi.org/10.1007/978-3-032-19288-2_5

Questions about the proper scale, scope, and character of matter have assumed a renewed sense of urgency given the emergence of the Anthropocene, a distinct geological epoch in which human activity has become so influential as to alter fundamental aspects of the Earth System. While ecologist Eugene Stoermer and Nobel Prize-winning atmospheric chemist Paul Crutzen introduced the current definition of the term in the 1980s, we have since witnessed a growing scientific consensus about the rigor of the concept. As an article published in the journal *Science* by the Anthropocene Working Group demonstrates, fluctuations in the content and pace of sediment deposits and extinction rates are unmistakably anthropogenically driven.[2] At the same time, the very nomenclature of the Anthropocene has been subject to critique from within the humanities for allowing an abstract notion of the Anthropos to anchor an implicit philosophy of history. Daniel Hartley notes that "inherent to the Anthropocene discourse is a conception of historical causality which is purely mechanical: a one-on-one billiard ball model of technological invention and historical effect, which is simply inadequate to explain actual social and relational modes of historical causation."[3] Hartley takes special issue with the presumed origins of the Anthropocene, which many geologists date to the industrial and nuclear revolutions. This determination, he suggests, interprets the environmental impact of technology as the "net effect" of an undifferentiated "human" activity.[4] In order to assert a causal link between technological development and ecological catastrophe, any consideration of the roles race, class, and gender have played in engineering our historical present must be obscured. The benefits and consequences of technological development and environmental disaster, after all, are rarely if ever distributed symmetrically. "It is not all people that are indicted by the onset of the Anthropocene," writes Nicholas Mirzoeff, "but a specific set: colonial settlers, enslavers, and would-be imperialists."[5]

This chapter explores how the concept of the Anthropocene further mystifies the ecological implications of the Middle Passage and the slave ship. By widening the focus of the preceding chapters to consider the "nonhuman turn" and the new materialisms in particular, I illustrate how the Anthropocene overdetermines our intellectual and ethical responses to environmental precarity. Many new materialist publications draw upon the techno-scientific advancements of the Anthropocene to demonstrate the supposed inadequacy of race-based politics for meeting the challenges of our time. But as a close reading of Octavia Butler's Afrofuturist novels

Parable of the Sower (hereafter *Sower*) and *Parable of the Talents* (hereafter *Talents*) reveals, any ethical model that dismisses considerations of race, and the history of transatlantic slavery specifically, will fail to address our environmental predicaments. I argue that by reimagining the slave ship as a spaceship, the novels stage a symbolic insurrection through which the ecology of the slave ship can be challenged. As I show in the previous chapter, this ecology is immanent to our environmental imaginaries and emerges as a tension between the formal and narrative elements of a creative work. Within Butler's duology, this tension manifests between its putative rejection of the black/white binary and its drive to return to the site of the binary's creation. Modern racial difference was forged fundamentally in the spacetime of the slave ship and the novels' goal of interstellar travel is achieved essentially through the remaking of that spacetime. In this regard, the trope of suspended animation stands in for the slave ship's regulation of environmental experience, sensation, and resource allocation. By commandeering and communalizing the environment of the ship, *Sower* and *Talents* attempt to disrupt the racial coordinates upon which the ecology of the slave ship and the Anthropocene are built.

The Anthropocene's scientific parameters may have been settled more recently, but its constitutive concerns—global warming, biotechnology, biodiversity loss, environmental racism—have thrown our prevailing conceptions of nature and culture into crisis well before the epoch's formal identification. At stake is not only the fate of *homo sapiens* as a species, but also the basic composition of a world yet to come. The challenges of analyzing the effects of nonhuman systems (e.g., weather patterns or ocean currents) and actors (e.g., viruses or pesticides) while attending to the uneven distribution of environmental risks and resources have generated a range of philosophical responses. For example, publications like Ursula Heise's *Imagining Extinction* and Elizabeth Callaway's *Eden's Endemics* probe the connections between biodiversity, conservation, and environmental storytelling. Other responses like Anna Tsing et al.'s *Field Guide to the Patchy Anthropocene* and Max Liboiron's *Pollution is Colonialism* interrogate fantasies of human mastery as a way of reckoning with the power of nonhuman agents. Over the last two decades, one particular response has acquired critical purchase within the academic left: the new materialisms.

As part of what Richard Grusin has labeled "the nonhuman turn," the new materialisms join affect theory, animal studies, and the blue humanities in calling for enhanced attention to matter and materiality. The popularity of this approach appears grounded in the need to cultivate strategies of coexistence attuned to the Anthropocene's political and ecological crises. How, for example, should we understand agency and embodiment in light of recent developments in biotechnology and the increasingly unpredictable behavior of nonhuman objects? What can these objects tell us about the relationships between humans and their environments? The promise of the new materialisms thus inheres in the notion that a focus on materiality can offer us more comprehensive and efficacious ways to respond to our changing circumstances. As Diana Coole and Samantha Frost write in their introduction to the *New Materialisms* anthology: "What is at stake here is nothing less than a challenge to some of the most basic assumptions that have underpinned the modern world, including its normative sense of the human and its beliefs about human agency, but also regarding its material practices such as the ways we labor on, exploit, and interact with nature."[6] There is much to recommend an intensified engagement with matter, not least of which is Coole and Frost's proposal that such engagements can disrupt our "normative sense of the human" and of "human agency."

Given this professed interest in dismantling human exceptionalism, it is curious then that as Zakiyyah Jackson and other critical race scholars point out, the new materialisms have systematically "[ignored] praxes of humanity and critiques produced by black people, particularly those praxes which are irreverent to the normative production of 'the human' or illegible from within the terms of its logic."[7] Indeed, as I suggest in previous chapters, the Middle Passage, the slave ship, and the transatlantic slave trade are responsible for modern understandings of the human and "objecthood." From at least the sixteenth century onward, black bodies provided crucial raw material for the development of natural history, the natural sciences, and the life philosophies in Enlightenment thought. Both geology and biology, for example, pursued notions of species and evolution that preserved early racial taxonomies; the techniques of observation and interpretation used to analyze geological activity were the same as those employed by the racial science of phrenology. Mirzoeff leverages this history to argue that "the very concept of observable breaks between geological eras in general and the definition of the Anthropocene in particular is inextricably intermingled with the belief in distinct races of humanity."[8] His claim that the conception of the Anthropocene reproduces race-making technologies gestures to the historical fact that the human as such has emerged through the exclusion and extermination of black bodies.

Proscribed from the realm of the human, black intellectuals have had to think within and through the categories of the nonhuman and the inhuman to pursue new ways of being in the world. Philosophical questions about the vitality and agency of the human, the animal, and the object are therefore long-standing in the fields of black studies. Alexander Weheliye observes that across Sylvia Wynter's oeuvre, "it is the human—or different genres of the human—that materializes as the object of knowledge in the conceptual mirror of black studies."[9] The scholarly work of Joy James, Ruha Benjamin, and Alexis Pauline Gumbs, to name but a few examples, similarly confront the "most basic assumptions that have underpinned the modern world," including our notions of history, temporality, and modern science. And yet, as it is with the Anthropocene's implied philosophy of history, much of the scholarship produced under the banner of the new materialisms tends to reduce race to a crude "identity politics" or to endorse as radical a model of difference-without-race.[10] This reduction and disavowal of race, I contend, is something of a structural necessity for the new materialisms.

In what follows, I trace the general theoretical principles of the new materialisms to a dissatisfaction with the linguistic and cultural paradigms of post-structuralism. I then demonstrate how this dissatisfaction enables an ethics of relation, or an ecology of difference, that further legitimizes the reduction and dismissal of race. However, as a close reading of Butler's *Parable* duology indicates, one of the primary figures of the new materialisms—the material body—is defined by and through disavowed social fantasies about black female flesh that are linked to the disciplinary formations initiated aboard the slave ship. My examination of the critical responses to Butler's novels further suggests that such fantasies are necessary to secure a libidinal investment in the ethical potential of materiality. I argue, thus, against a misrecognition of black female flesh as a resource against the violence of hierarchical differences, rather than the site of their active production. Finally, I turn to a reading of Butler's *Parable* duology as an allegory about the dangers of proceeding in the Anthropocene without a robust analysis of race and racial blackness. In doing so, I demonstrate how the novels' use of suspended animation clarifies further the ecology of the slave ship and its underwriting of the Anthropocene.

The Promise of the New Materialisms

The new materialisms are drawn from a long genealogy of philosophical materialism, in which Spinoza, Nietzsche, Bergson, Marx, and Deleuze are cited as major touchstones. In recognition of this legacy, Coole and Frost assert that the interventions loosely gathered by the term "new materialisms" are better "categorized as *renewed* materialisms," with the qualifier "new" acknowledging the "unprecedented" ecological, biological, and technological conditions under which we currently live and labor.[11] Although their specific objects of analysis are appropriately diverse, the new materialisms collectively insist on a posthumanist matter that is lively, self-directed, agential, creative, and always in the process of becoming. In this regard, matter is better thought of as *materialization*, or the process by which complex phenomena are temporarily and contingently stabilized to varying degrees. The ontological shift entailed here is toward a philosophical monism, inspired most notably by the work of Deleuze. Following Spinoza and Bergson, Deleuze develops a notion of the virtual as a generative field of difference, or a "plane of immanence," where "all the varieties of differential relations and all the distributions of singular points [coexist] in diverse orders 'perplicated' in one another."[12] These differences are then formatted into distinct phenomena or entities by processes of actualization that "[bring] the object back into relation with the field of differential relations in which it can always be dissolved and become actualized otherwise, as something else, by being linked through other differential relations to other particles."[13] While not all new materialist theories cleave to a strictly Deleuzian philosophy, there is general agreement that the interactions among objects, bodies, and phenomena turn us away from the Anthropocene's "billiard ball model" of causality, and more significantly, from some of post-structuralism's critical trends.

According to the new materialisms, the linguistic and cultural turns of the last half century have resulted in both an intellectual and political poverty. Specifically, social constructivism and cultural representationalism have overdetermined matter to the extent that it appears as a passive product made meaningful only through cultural and discursive practice. Coole and Frost even write of a theoretical "exhaustion," claiming that they "share the feeling current among many researchers that the dominant constructivist orientation to social analysis is inadequate for thinking about matter, materiality, and politics in ways that do justice to the contemporary context of biopolitics and global political economy."[14] Somewhere and sometime during the rise of the Anthropocene, cultural theory, broadly conceived, lost its explanatory power. This assessment of inadequacy repeats across much of the recent new materialist scholarship, condensing the cultural turn into a discursive reductionism that rebuffs the empirical for the ideal, or the material for the symbolic. Elizabeth Grosz's *The Nick of Time* opens with a telling.

> reminder to social, political, and cultural theorists, particularly those interested in feminism, antiracism, and questions of the politics of globalization, that they have forgotten a crucial dimension of research...not just the body, but that which makes it possible and which limits its actions: the precarious, accidental, contingent, expedient, striving, dynamic status of life in a messy, complicated, resistant, brute world of materiality.[15]

Social, political, and cultural theory, in other words, has overlooked the material conditions of life that leave the body available for inscription and enculturation in the first instance. So too in the recently published *Gut Feminism* does Elizabeth Wilson rebuke "social constructionism" for "[tending] not to be very curious about the details of empirical claims in genetics, neurophysiology, evolutionary biology, pharmacology or biochemistry."[16] Her ensuing conclusion is that focusing on how social structures produce and discipline bodies comes at the expense of recognizing the ways bodies radically alter and organize social structures themselves. It appears that cultural theory harbors an "allergy to 'the real'" that dissuades "critical inquirers from the more empirical kinds of investigation that material processes and structures require."[17] However, the very aspects that would make matter more "real" than language or culture are the same aspects that restrict its ethical potential and facilitate a conceptual rejection of race.

In line with a posthumanist agenda, the new materialisms evoke an ecology of difference, in which matter and materiality exist in excess of human subjectivity. Mechanistic theories of causality hold that objects are composed of inert matter acted upon by external forces, which presumes that an object's potential or possible capabilities are already present and fixed in some initial moment of creation. But, as the new materialisms emphasize, the virtual field of differential relations is immanent to matter in such a way that it is impossible to anticipate all of the effects a material configuration may have, or the forms it may take. This ability to act independently of the human subject's will and desire is variously construed as "something that is both more and other than that which I think of as me and mine," an alterity that "comes from outside the capability or power of the subject," "degrees of indetermination" that represent the "true principle of life," and a "powerful reminder…that life will always exceed our knowledge and control."[18] Differences in terminology aside, the new materialisms are united by an understanding of matter as a spectral, impersonal force; one that escapes reason and transverses systems of meaning, including modernist binaries like mind/body, nature/culture, and inside/outside. The latter aspect is key because while matter can frustrate representation, its "excessive" properties do not mean that it exists "outside" of the subject. Rather, matter and materiality are "real" because they actively produce reality in unpredictable ways. It is here that the ethical impetus of the new materialist project is located.

New materialist ethics necessarily manifest as affective encounters that operate best on micropolitical scales. Because materiality is figured as an impersonal force of the real, it runs the risk of becoming a transcendental signified that merely replaces language or culture as an organizing principle. Doing so would severely diminish its import as an inducement to a posthuman ethics. To circumvent the "tension between universalistic theory and specific mode of inquiry," chance, contingency, and creativity in micro-level encounters are prioritized over more obstinate assemblages that congeal at the global or macro-levels.[19] Further, as the nucleus of the new materialisms, the embodied subject or material body compels an ethics that unfolds on a parallel plane, meaning between and within bodies. "This implies," Rosi Braidotti proposes, "approaching the world through affectivity and not cognition: as singularity, force, movement, through assemblages or webs of interconnections with all that lives," and "accepting the impossibility of mutual recognition and replacing it with one of mutual specification and mutual codependence."[20] If we accept our embeddedness in these mutually transformative, nonhuman ecologies, the ground of ethics shifts accordingly. First, a responsibility to an externalized other gives way to an accountability for the many relations that constitute becoming. And second, ethics are no longer reducible to the decisions or actions of individuals that are initiated by a properly historical judgment. Though I find nothing immediately problematic with an ethics that includes nonhuman and inorganic phenomena and objects, what I do find troublesome is how our acquiescence to these ethics is solicited.

In the quotation above, Braidotti invokes an ethics of relation, in which sensation and perception comprise the "zone of [ethical] effectivity," and attunement and affirmation take precedence over social transformation.[21] Because material inter- and intra-actions are preconscious and multisensorial, ethical practice is based not on the ability to evaluate right from wrong, but on a commitment to *feeling* right. We can observe this adjustment in appeals to "an ongoing responsiveness to...entanglement," "a heightened sensitivity to the agency of assemblages," a "wakefulness" to the "feel [of] what makes us laugh, lament, and curse," and an "experience of the vitality of being."[22] As a consequence, the experiences of living under conditions of crisis are fetishized at the expense of addressing the causes of these conditions themselves. The imperative to "[live] with the open wound...through a sort of depersonalization of the event," for example, not only depoliticizes the claims of historically oppressed communities, but also flattens distinctions between traumas inflicted through happenstance and persistent intergenerational *harm*.[23] How else could Braidotti list as equivalent examples: those who survived the Holocaust, Frida Kahlo's deadly tram ride, and missing the train to the World Trade Center on September 11th?[24]

The limits of a new materialist ethics appear most forcefully, then, as we attempt to move from an embodied "responsiveness" to the dislocation of structures. When patterns of materialization are addressed, it is generally as the amalgamation of "perpetual circuits of exchange, feedback, and reentry" that thereby "[inflect] the shape of political experience."[25] On the one hand, there is nothing innately objectionable about attributing the creation and transformation of political structures to any number of quotidian, embodied experiences. This is in fact common in political theory and historiography.[26] On the other hand, it becomes more difficult to reconcile the effects of chance, unpredictability, and indeterminacy with the endurance and repetition of something like antiblack violence.[27] The new materialisms are therefore at pains to clarify why the structures of global antiblackness continue to function as if "neither time nor history, nor historiography and its topics, show movement, as the human subject is 'murdered' over and over again by the passions of a bloodless and anonymous archaism, showing itself in endless disguise."[28] Interpreting and describing our entanglements with nonhuman, materialist forces are not enough to account for, much less dislodge attachments to, social categories and representational arrangements. By this I mean that becoming more aware of material forces will not inevitably reduce the weight of discursive or psychic formations. It could even obstruct change by making forms of affect and sensation newly available for inscription. As Timothy Morton states, when "contact becomes content," perceptions of difference collapse into identity.[29] Granted, these complications are not unique to the new materialisms as changes in scale almost always entail a re-calibration of ethics. The point is, however, that positing an ecology of difference as the ethical ground of the Anthropocene requires a disavowal of race and its material-semiotic effects.

As I submit earlier in this chapter, since at least the Enlightenment period intellectual genealogies have maintained an almost overwhelming racial homogeneity. Critical theories produced by non-white scholars certainly existed but were and still are marked as minority perspectives that had little to do with questions of a universal or ontological character.[30] Hence, black bodies especially are rendered objects for theoretical development, rather than subjects of universal philosophy. Coole and Frost continue this trend, commenting that even as "feminists and class theorists have often insisted upon" the importance of material bodies and environments, they remain "[concerned] that such material dimensions have recently been marginalized by fashionable constructivist approaches and identity politics. Of course, the latter have had a good deal to say about the body and its imbrication in relationships of power, but we are not convinced that they pay sufficient attention to the material efficacy of bodies or have the theoretical resources to do so."[31] Such a statement is heavy with long-standing racial charges of intellectual primitivism and parochialism. Their unfortunate request to be "convinced" of identity politics' intellectual merit effectively seals an historically white critical theory as the standard for authoritative knowledge production. One must also wonder about the referents for these insufficiently materialist "identity politics," given that the *New Materialisms* anthology fails to cite even one example that might be taken as representative of a larger trend. Even if Coole and Frost employ "identity politics" as a shorthand for idealist approaches to subjectivity, their statement at best misunderstands studies of "identity," and at worst discloses a symptomatic desire to abandon race. To be clear, Coole and Frost never openly reduce "identity politics" to racial identity. But in many if not most of new materialisms' founding texts, race receives only casual mention alongside the "other so-called axes of social difference" like sex, gender, and class, and often only to specify a concept that has been "paralyzed by [a] 'binary' take on dualism," or to designate potential beneficiaries of one's theorization.[32] We could perhaps attribute this treatment of race to a sedimented politics of attention that determines which issues receive scholarly consideration.[33] Nonetheless, to ascertain if and how the new materialisms might furnish us with a timelier ethics, we must first ask what purpose the disavowal of race serves.

The Black Lives Matter movement offers a powerful reminder that black bodies have historically provided the standards against which the human and nature are measured. This is to say that the "rupture in the quality of being" inaugurated by modern racial slavery is not limited to black lives.[34] Black critical theorists repeatedly insist on the world-historical scale of this rupture, tracking how it fashions our thinking about nature and the human, and the movements of this thought itself. What this means for our current discussion is that "the question of race's reality has and continues to bear directly on hierarchies of knowledge pertaining to the nature of reality" or on what Dionne Brand calls our "cognitive schema."[35] As a conceptual orientation or method of "way-finding," the prevailing cognitive schema articulates a libidinal economy of antiblackness to the history of ideas, ensuring, as Hortense Spillers maintains, that "dominant symbolic activity, the ruling episteme that releases the dynamics of naming and valuation, remains grounded in the originating metaphors of captivity and mutilation"; after all, "sticks and bricks might break our bones, but words will most certainly kill us."[36] By inverting a youthful rejoinder about the supposedly limited reach of the symbolic, Spillers lays out a provocative proposal: the metaphors of slavery are immanent to the force of materiality.[37] Although "'race' alone bears no inherent meaning, even though it reifies in personality," it "gains its power from what it signifies by point, in what it allows to come to meaning."[38] Black lives matter and blackness enlivens matter. It is possible, then, that the elaboration of thought and the conditions of its enunciation are always part of a racial praxis, even when those "personalities" that expedite the reification of race are most absent. This is a paradigmatic example of the prevailing cognitive schema at work. Antiblackness conditions the force of materiality by determining the logic and legibility of its actualization. These functions become clearer when we turn our attention to Octavia Butler's *Parable* duology.

Parables for Our Time

The *Parable* novels are set in a dystopian America, produced by a fifteen-year period of "coinciding climactic, economic, and sociological crises" known collectively as "The Pox."[39] Amid the ongoing economic and political collapse of the United States and its de facto elimination of social services, protagonist Lauren Olamina lives with her family in the Southern California gated community of Robledo, which is Spanish for "Oakwood." The community's walls offer an illusion of security to the semi-professional, non-white residents, and despite the increasing violence just outside, they refuse to contemplate alternatives. Angered by this inability to concede the permanence of change, Lauren devises a line of flight in the form of a political theology she names "Earthseed." She becomes the titular "Sower" of the first novel, guided by Earthseed's central principle that "God is Change." After the eventual massacre of the people of Robledo, Lauren travels to Northern California with fellow itinerants she recruits along the way. *Parable of the Sower* closes with their founding of the town of Acorn, returning the overly mature and moribund idea of "the sturdy old oak" to its source, a community committed to Earthseed's doctrine of adaptability, self-sufficiency, and diversity. While both novels are narrated through Lauren's journal entries, *Parable of the Talents* includes contributions from Lauren's daughter, Asha, and Lauren's husband, Bankole. *Talents* picks up five years after the events of *Sower* and details the imprisonment and dispersal of the Acorn community by a fundamentalist Christian sect. In the second half of *Talents*, Lauren sets out to find a missing Asha while attempting to rebuild the Earthseed movement. The narrative culminates with the departure of the world's first interstellar spaceships, sponsored by the now powerful network of Earthseed acolytes.

What makes the novels imminently relevant to our discussion is that the apocalypse-inducing Pox begins in 2015. The formal narration commences in 2024, but the overlaps between these diegetic and extradiegetic levels endorse a reading of the novels as cautionary tales for the Anthropocene of our present. To be sure, the eponymous "parables" of the titles refer to the New Testament's instructional stories, and there is a great deal in Butler's novels deserving of our caution. All of the Anthropocene's most troubling possibilities are realized by 2024. The rapidly warming climate decimates crops and creates monster storms, displacing millions of people and exacerbating the spread of hunger, poverty, and disease. As US citizens attempt to flee to Alaska, "The Last Frontier" eventually secedes from the union to form a Northern Bloc with Russia and Canada. By 2033, large-scale warfare has broken out. As Butler makes clear, these events were set in motion prior to the onset of the Pox and considerably before 2015. This warp of time and history implicates our existing socioeconomic orders in a future that by most accounts is already part of our present. It is here that literary critics locate the basis for Lauren's political vision. Her postmodern embrace of change is designed to interrupt the immobilization of difference that produces all hierarchical systems, including those of race, class, gender, sexuality, and matter. According to Madhu Dubey, this productively sets the *Parable* novels apart from the speculative fictions of slavery that arose during the 1970s.

From the antebellum period to the Civil Rights era, African American literature functioned most notably as a means of social protest. Realist narratives of slavery and post-emancipation black life lent crucial support to political movements and supplemented the historical archives before substantial numbers of first-person slave testimonies were recovered. The post-Civil Rights period then witnessed a renewed literary interest in slavery as black communities sought innovative ways of understanding the past in relation to their ambiguous political futures. African American novelists began to turn away from realism and toward elements of the speculative and fantastic. Concurrent with a broader postmodern rejection of historiography, these speculative fictions of slavery "overtly situate themselves against history, suggesting that we can best comprehend the truth of slavery by abandoning historical modes of knowing."[40] Literary devices like time travel, gothic hauntings, and spiritual possession not only fold the present directly into the antebellum slavery of the past, but also depict slavery as a structure of feeling best understood through material experience. But portraying the past as continuous with the present, Dubey concludes, prevents readers from distinguishing "the pluralist racial dynamics of our present from the binary racial logic of antebellum slavery."[41] For her, the multiracial debt slavery of the *Parable* novels is a more accurate representation of those dynamics.

Instead of physically transporting her characters into the antebellum era, Butler conjures slavery in the present through and as its historical influences on an evolving socioeconomic and political order. The *Parable* world is consistent with the worldwide expansion of neoliberal ideology and practice. The cataclysmic events of the Pox not only exposed new segments of the US population to exploitation, but also extended the domain of what Naomi Klein calls "disaster capitalism."[42] That is, the growing scarcity of natural resources presents the perfect opportunity for their large-scale privatization. Multinational corporations like KSF, "Kagimoto, Stamm, Frampton: Japanese, German, Canadian," begin purchasing communities and small businesses to "dominate farming and the selling of water and solar and wind energy."[43] Residents and migrant laborers are promised security and stable employment in exchange for nominal wages, an "old company-town trick" that signals "something old and nasty" has been revived.[44] But unlike the company towns of the late nineteenth century, many of those created by conglomerates like KSF were once "upper middle class, white, literate [communities]."[45] Hence, the "slaves" of Butler's twenty-first century hail from all ethnic and racial groups and from populations formerly protected by wealth or whiteness. As Lauren conducts her party along the highway to the site of Acorn, she recognizes that they have "become the crew of a modern underground railroad."[46] What makes their flight "modern" is that the travelers of this Underground Railroad form a "heterogeneous mass – black and white, Asian and Latin."[47] As such, their journey to the north is just as likely to recall fugitive slave narratives as it is the contemporary movement of immigrants up from the Global South.[48] In addition, the types of slavery in the *Parable* novels range from indentured servitude and debt slavery to sexual trafficking and religious imprisonment.

The mechanisms of control necessary for US chattel slavery are retrofitted to meet the twenty-first century's new standards of precarity. Four of the founding members of the Acorn community, Emery, Tori, Grayson, and Doe, escape from similar situations before joining Lauren's group. Emery and Tori, who Lauren describes as "the most racially mixed [people]" that she has ever met, worked on a local farm with Emery's husband and their two sons before it and they were sold to an agricultural corporation.[49] By paying wages in "company script" and charging exorbitant prices for rent, food, household items, and clothing, the corporation ensured that workers were soon buried in debt.[50] The *de jure* and de facto repeal of labor laws likewise granted companies the right to obtain repayment in any way they saw fit, which included workers being "traded and sold with or without consent, with or without their families, to distant employers" and forcing children "to work off the debt of their parents if the parents died, became disabled, or escaped."[51] Accordingly, after the death of Emery's husband, her two sons are taken without warning and trafficked sexually. It is only after evading a litany of security and surveillance measures—sound and motion detectors, armed guards with dogs, and electric fences—that Emery and Tori manage to reach the highway. Grayson and his daughter Doe also join Lauren's group shortly after Grayson's wife is raped and killed by her new owner.

Within the logic of modern racial slavery, the separation of families, deployment of sexual violence, and enclosure of the commons were crucial to the maintenance of racial hierarchy. Conversely, their use in the *Parable* duology suggests that these practices have become disarticulated from the projects of race-making and are used instead to render both bodies and natural phenomena as abstract signifiers of value. The capitalist paradigms of exploitation and resource extraction, we are to presume, are the best ways to understand the distribution and arrangement of power. Perhaps unsurprisingly then, literary critics tend to minimize the significance of Lauren's racial blackness. Patricia Melzer provides a working example, writing in a passage worth quoting at length:

> Butler's approach to race issues that at first appear to be in the background of her social critique can be understood as a (narrative) strategy that undermines the binary of white/black that dominates U.S. discourse on race relations...Instead, Butler places racial oppression into the complexity of social power relations, such as in terms of economic and ideological oppression. She does not foreground racial oppression in her analysis of social injustice, but undermines the juxtapositions and binaries of racial discourse of self/other by portraying racial diversity as a main component of her utopia vision...Butler firmly roots her protagonist within an African-American context, yet at the same time she refuses to ideologically ghettoize her characters...By insisting on the presence of people of color in her narratives as normal, not exceptional, Butler also implicitly rejects the tokenism that categorizes her work primarily in terms of her identity as African American.[52]

As it is within the new materialisms, "race," which is really a placeholder for "the binary of white/black," is dissolved into a vague "tokenism" that is itself a stand-in for "identity politics." Melzer's arguments are part of a larger pattern in which literary critics position Butler as a science fiction writer who is black, rather than as a black woman who writes science fiction. This may seem peculiar, given that Butler's work regularly employs narrative strategies and themes consistent with black literature and experience (e.g., *Kindred*, the *Lilith's Brood* trilogy, and *Fledgling*). After the publication of her *Patternist* series in the early 1980s, she was hailed as a welcome corrective to science fiction's largely white, colonial, and patriarchal overtones.[53] But for her part, Butler has expressed in interviews that "it is a writer's duty to write about human differences, all human differences, and help make them acceptable."[54] Gregory Hampton draws out the implications of this undertaking, declaring that "Butler's fiction is successful largely because it produces narratives that are easily comparable to African American experiences but also because it considers the perspective of a universal marginalized body."[55] If we follow the lead of critics like Dubey and Meltzer, or the new materialist refusal of "identity politics," the "universal marginalized body" manifests as a body suspended in a static nexus of identity and representation, which is to say that all bodies, to varying degrees and at different times, are marginalized bodies. What we lose in this rush from the particular to the universal is any consideration of how the material-semiotic history of race governs, from the outset, what can and cannot be made legible as a universal. In critical treatments of the *Parable* novels, slavery is only admitted to the status of the universal as a set of exploitative and extractivist strategies, but rarely as a material force that lingers in the body to reproduce and confirm the signifiers of race. Taking these analytical gaps as my point of departure, I ask: how might our theorizations of matter and materiality transform if we understand slavery as a material force that works through the body and between the body and its environment?

Hyperempathy and Body Knowledge

One of Butler's most consistent themes is an embodied mode of knowing and communicating that she calls "body knowledge." In accordance with her belief that "all we really *know* that we have is the flesh," her narratives are centered on the body as the primary location for the disassembly and reassembly of regimes of power.[56] Our sense of our bodies is coordinated through these regimes to varying degrees, most of which suppress or otherwise manipulate body knowledge in service of their own maintenance. In the wake of eugenics, scientific racism, and the retrenchment of reproductive rights, it is difficult to imagine a non-neurotic relation to the body. This, Butler proposes, should not discourage us from learning how to reclaim and repurpose body knowledge for our own survival. The *Parable* novels take on this task by amplifying the intimate processes through which body knowledge is acquired. Lauren suffers from a condition called hyperempathy, an "organic delusional syndrome" that obliges her to share other people's pain and pleasure.[57] This heightened sensitivity is visually activated; in her own words, "I feel what I see others feeling or what I believe they feel."[58] Despite being imaginary, Lauren's ability to "share" leaves her dangerously vulnerable to the violence that permeates both novels. Speaking to Bankole, she laments, "Self-defense shouldn't have to be an agony or a killing or both. I can be crippled by the pain of a wounded person. I'm a very good shot because I've never felt that I could afford to just wound someone.... The worst of it is, if you got hurt, I might not be able to help you. I might be crippled by your injury – by your pain, I mean – as you are."[59] As a consequence, Lauren's social interactions are governed by a kind of strategic calculus; the "Golden Rule" enacted not just at the level of bodies but between them.

For many critics, hyperempathy encourages more ethical approaches to difference by releasing the body from its historically fixed positions. Variously described as the "right medicine for our present 'compassion fatigue,'" "the living embodiment of the subversion of difference," and "a crucial metaphor for re-defining social relations," hyperempathy seems to carry an almost utopian ethical potential.[60] In this regard, hyperempathy accomplishes what E.P. Thompson identifies as the pedagogical function of utopia, or "the education of desire."[61] The scholarly reception of the Parable duology announces, therefore, that a radical break with our attachments to the body is not only possible but also desirable. We are now firmly within the scope of the posthuman and nonhuman turns. The body hailed in the *Parable* duology is precisely the body theorized in the new materialisms. For instance, Coole and Frost define bodies as "open series of capacities or potencies that emerge hazardously and ambiguously within a multitude of organic and social processes," indicating that the materiality of the body exceeds whatever provisionally coherent and stable form it may take.[62] The material body is a temporary albeit stubborn configuration of a deeper flow of difference. To illustrate, after experiencing several incidents of sharing pain, Lauren notes: "I had no sense of my own body. I hurt, but I couldn't have said where – or even whether the pain was mine or someone else's. The pain was intense, yet [diffuse] somehow. I felt…disembodied."[63] Lauren's inability to distinguish between her pain and "someone else's" prevents her from locating her body in space and time, demonstrating that the body's inhabitation of "an Umwelt that remains ambiguous, indeterminate, and resonant with an expressive Significance…affects the body's perception of spatial relations."[64] This loss of proprioception, or the ability to make immediate "sense of" her body, suggests not only that bodies can act and respond prior to rational cognition, but also that the body is a non-deterministic form produced through and traversed by the "open series of capacities or potencies" foregrounded in Coole and Frost's definition.

Hyperempathy also shares with new materialist philosophies a capacity to upend the social and political hierarchies that regulate our encounters with difference. As literary critic Jerry Phillips agrees, "In a hyperempathetic world, the other would cease to exist as the ontological antithesis of the self, but would instead become a real aspect of oneself."[65] For one, hyperempathy is not limited to connections between human bodies. Before the destruction of Robledo, Lauren and a group from the community venture beyond the walls to hunt for her brother. After her father shoots a feral dog, Lauren realizes that it is still alive: "I saw its bloody wounds as it twisted. I bit my tongue as the pain I knew it must feel became my pain…With my right hand, I drew the Smith & Wesson, aimed, and shot the beautiful dog through its head…I walked, then rode in a daze, still not quite free of the dog I had killed."[66] For Lauren, the dog's death lingers as the ethical ramifications of violence are translated into an exchange of affect between bodies. Her killing of the dog is as much an act of mercy for it as it is for her, which seems to confirm Braidotti's supposition that "affectivity in fact is what activates an embodied subject, empowering him or her to interact with others…it follows that a subject can think/understand/do/become no more than what he or she can take or sustain within his or her embodied, spatiotemporal coordinates."[67] That one of these bodies is coded as nonhuman or animal has no purchase on its ability to act as a causal force. Without a doubt, differences exist between Lauren and the dog but at the level of the materialist body these differences cannot cohere into social categories.

Additionally, Lauren's experiences with difference resolve into a central tenet of Earthseed: "Embrace diversity/Unite–/Or be divided,/robbed,/ruled,/killed/By those who see you as prey./Embrace diversity/Or be destroyed."[68] The implied choice is between two perspectives on difference: the first—diversity—views difference as the foundation for collective empowerment, while in the second difference continues to scaffold a social hierarchy in which some people emerge as prey. The events of the Pox have created an economy of survival, where "people are expected to fear and hate everyone but their own kind."[69] In defiance of this view, Lauren saves and then invites a young family to join her group, commenting, "We're natural allies - the mixed couple and the mixed group."[70] Her commitment to diversity prompts Melzer to contend that hyperempathy yields a "shared identity and life experience that [is] not based in a particular unified racial or cultural background," and Phillips to declare that Butler "employs a race-transcendent communalist ethics."[71] In other words, for the material and hyperempathic body to serve the human community, it must be free of race.

An anxiety that race might follow this body is well founded, given how Butler remarks that hyperempaths make good slaves. After learning of her brother's horrific death, Lauren asks: "…if everyone could feel everyone else's pain, who would torture? Who would cause unnecessary pain? I've never thought of my problem as something that might do some good before, but the way things are, I think it would help."[72] Nonetheless, in a 2001 interview with NPR, Butler explicitly argues that "the threat of shared pain wouldn't necessarily make people behave better toward one another."[73] The proliferation of human trafficking and indentured servitude makes this clear as company "bosses" or "drivers" pay extra for workers with hyperempathy syndrome.[74] Moreover, hyperempathy syndrome is induced in utero by the maternal use of Paracetco, a drug designed to impede the degeneration of brain functions in Alzheimer's disease. Paracetco also increased the intellectual capabilities of non-afflicted users and so became the stimulant of choice for the middle class, which included Lauren's mother. And like the antebellum law of *partus sequitur ventrem* that mandated children follow the status of their mothers, the children of hyperempaths can also inherit the condition, even if they are several generations removed from the original drug use. Despite these connections to racial slavery, hyperempathy, like the debt slavery of the twenty-first century, is not racially exclusive. And yet, the relationships between hyperempathy and a slavery of the future are telling insofar as they bring the specter of blackness back to the fore.

What Lauren's hyperempathy elucidates is that in order to free difference and the body from their historical constraints, we must attend to one particular constraint by which they are bound—racial blackness. This may seem paradoxical when we consider that new materialist scholarship almost uniformly disregards race in its return to the body. But the *pro forma* rejection of race as ensnared in either identity politics (i.e., the new materialisms), or an "obsolete" black/white binary (i.e., literary criticism), is more correctly a disavowal that untethers the non-, in-, and posthuman from their historical site of production. More precisely, disavowing the associations between the non-, in-, and posthuman and racial blackness defends theories of the former against the ontological and conceptual provocations of the latter. The proscription of blackness in the new materialisms and in the responses to the *Parable* duology attests to this fact, even as they establish blackness as the vanishing mediator between the Anthropocene and its possible futures. Indeed, as Butler's novels show, the matter and materiality of the Anthropocene often look and feel like the historical traumas of slavery.

Suspended in the Flesh

In the Afrofuturist context of the duology, the slave ship is reimagined as a spaceship with the hold becoming a site of suspended animation.[75] In response to several world-alerting events, Earthseed attempts to give humanity a "unifying, purposeful life" by taking to the stars.[76] The hope of a "real heaven, not mythology or philosophy," Lauren believes, will motivate humanity to perceive itself as "a growing, purposeful species" made up of more than "smooth dinosaurs who evolve, specialize, and die."[77] These long-range, interstellar missions require substantial advances in science and technology and as a testament to Earthseed's motivational strengths (and fund-raising capabilities), the first shuttles leave earth for a starship at the end of *Talents*:

> Today's shuttles have been loaded with cargoes of people, already deeply asleep in DiaPause – the suspended-animation process that seems to be the best of the bunch. Traveling with the people are frozen human and animal embryos, plant seeds, tools, equipment, memories, dreams, and hopes. As big and as space-worthy as they are, the shuttles should sag to the Earth under such a load. The memories alone should overload them. The libraries of the Earth go with them. All this is to be off-loaded on the Earth's first starship, the *Christopher Columbus*. I object to the name. This ship is not about a shortcut to riches and empire. It's not about snatching up slaves and gold and presenting them to some European monarch.[78]

In a reversal of the colonial and science fiction trope of "first contact," the *Christopher Columbus* is traveling beyond the known galaxy neither for "riches and empire" nor for "slaves and gold," but to "take root among the stars."[79] The word "cargo" nods to Lauren's earlier description of the ships as "space trucks" and to be sure these interstellar moving vans are laden with the knowledge, tools, and materials necessary for establishing homes on other planets.[80] At the same time, because the living members of this collective are "frozen" or "deeply asleep," the word "cargo" signals a second reversal linked to the slave ship. With this multispecies cargo already in suspended animation, the hold becomes a space that preserves or extends their "memories, dreams, and hopes" rather than dismantling them. Whatever distinctions exist between the hold and the deck are functional instead of ontological. To be clear, there is nothing unusual about Butler's recourse to suspended animation. To move characters across time and space without exposing them to the effects of entropy, science and speculative fiction writers turn regularly to this narrative device. What makes it notable to and within these novels is its relationship to the duology's other psychosomatic technologies—hyperempathy and the slave collar. In this register, suspended animation appears as an attempt to renegotiate the ecology of the slave ship and the racial dynamics it authorized.

In the preceding chapters, I argue that the slave ship established a network of environmental relations that helped to make and manage racial difference. By regulating the slave's experiences of and relationships to their environments, sailors and non-black authority figures were able to naturalize the effects of racial violence while extracting value from its application. Earthseed's introduction of suspended animation supplements this template by guiding our attention to the slave's sensorial array. In effect, the slave ship supplied the prototype for suspended animation. If we think of suspended animation not as a state of passivity or stasis but as the regulation of the ability to detect, contextualize, affirm, and otherwise process environmental sensations, then we are closer to understanding how the ecology of the slave ship continues to manage environmental experience. The forces of materiality may arise without human assistance, but we are more than capable of manipulating when, where, and how those forces are apprehended. Suspended animation, moreover, bears more than a passing resemblance to what Hortense Spillers calls "flesh." In her essay, "Mama's Baby, Papa's Maybe: An American Grammar Book," Spillers traces the construction of the modern body and its allegorical possibilities to the violences of transatlantic slavery. The reduction of the African captive to "a thing" was accomplished through a concomitant severing of the body from "its motive will [and] its active desire."[81] This "theft of the body," Spillers writes, requires that we make "a distinction in this case between 'body' and 'flesh' and impose that distinction as the central one between captive and liberated subject-positions…before the 'body' there is the 'flesh,' that zero degree of social conceptualization that does not escape concealment under the brush of discourse, or the reflexes of iconography."[82] In registering the wounds of slavery, "flesh" is both a site of physical injury and a "primary narrative" from which the signifiers of subjectivity are painfully extracted. So too, suspended animation is simultaneously a material and metaphorical state of being that hinges on the distinction between body and flesh. Or more specifically, in seizing the slave's means of sensation, the body is suspended in favor of the flesh.

Near the beginning of *Talents*, a newly elected fundamentalist and fascist government sends its agents to invade Acorn and imprison its residents. Acorn, refashioned as the "Camp Christian Reeducation Facility," becomes host to almost 250 prisoners, most of whom are incarcerated for being poor or having the wrong religious affiliation. During their captivity, inmates are fitted with a control device known as a "slave collar," which can deliver pain or pleasure at the discretion of its controller or whenever the collar or its control unit is tampered with. Unlike hyperempathy, which is based on shared sensory experience, the collars operate through unilateral sensory manipulation:

> some collars [can encourage] changes in brain chemistry…the whole business sounds a little like being a sharer – except that…the wearer feels whatever the person holding the control unit wants him to feel. This could initiate a whole new level of slavery. After a while, needing the pleasure, feeling the pain, and always being desperate to please the master could become a person's whole life. I've heard that some collared people kill themselves, not because they can't stand the pain, but because they can't stand the degree of slavishness to which they find themselves descending.[83]

In describing the slave collar as a structure of force that regulates sensation, Lauren also reframes "slavishness" as a state of suspended animation, one that occurs when the slave's flesh is aligned with the master's will and desire. Flesh becomes the medium through which the master's wishes are translated into a sensory language of pain, pleasure, mood, and affect. And because the collar enforces a system of reward and punishment through this language, the slave must develop a repertoire of behaviors in relation to the master's needs. Lauren's observation that the collar "makes you turn traitor against your kind, against your freedom, against yourself" is verified when two Acorn community members report a forbidden queer relationship to their captors, who "lash" the offending couple until one of them dies.[84] Some non-Acorn prisoners also become informants, alerting the guards to infractions to receive better treatment or to avoid punishment.

In this way, the slave collar is a reification of Sylvia Wynter's "sociogenic principle," or a "culturally imposed symbolic belief system" that "serves as the internalized sanction system that motivates behavior, biochemically affirming or negating in dynamic relation to societal norms and values prior to any reflective process. A species-specific opioid (reward and punishment) system serves to induce its appropriate behaviors through the mediation of each person's subjective experience of what feels good and what feels bad to and for each person."[85] Like the slave ship, the slave collar translates practices of race-making into bio-sensory information that guides a captive's behavior, body knowledge, and sense of self. At the same time, its "severing" of the captive's body from their will and desire is not equivalent to the erasure of that will and desire. While discussing her "modern underground railroad," Lauren muses, "if we can convince ex-slaves that they can have freedom with us, no one will fight harder to keep it."[86] Her supposition turns out to be true when Emery and Grayson are killed during the invasion of Acorn, both choosing death over re-enslavement. Other Acorn members also commit acts of resistance that stretch from whispered conversations and illicit intimate encounters to open revolt and murder. To an extent, these acts point to the failure of the collars to colonize fully the inner lives of their captives. What the aforementioned severing does achieve is the transformation of the slave's flesh into a "territory of cultural and political maneuver."[87] The slave collars are instrumental to Camp Christian's proselytizing mission, as they facilitate the conversion of the slave's flesh into the body of the Christian. By arresting the slave's body knowledge, the collars train the flesh to associate specific sensations with specific behaviors: "Everyone...worked for sixteen hours straight. They lashed you if you stopped to pee...dig a hole. Fill it up. Chop trees. Make firewood. Dig another hole. Fill it up."[88] In this way, the Camp gains several informants who "started to believe all those sermons and Bible classes and prayer meetings and the other stuff [the guards] made [them] sit through or stand through when [they] were almost too tired to live."[89]

While the practice of breaking the body to subjugate the mind has a long history in institutions like works camps and the military, the slave collar increases the efficiency of the practice by internalizing its tactics. In this respect, the collar speaks to contemporary fears over the power and potential of Anthropocene technologies to interrupt self-determination on a biochemical level. Lauren expresses a similar fear after a severe lashing causes her to forget the death of Bankole: "Being lashed hard several times is enough to induce temporary coordination problems and memory loss…After surviving Robledo, I knew that strangers could appear and steal or destroy everything and everyone I loved. People and possessions could be snatched away. But somehow, it had not occurred to me that…that *bits of my own mind* could be snatched away too."[90] Her memory loss proves to be temporary but as indicated by the words "steal" and "snatched away," the manner in which it was lost encouraged a proprietary orientation toward her mind and self. The slave collar makes material the conflation of personhood and property that was established through racial slavery and reinforced in its afterlife. In doing so, it threatens not only to reinforce the equation of self-determination and self-ownership but also to reinstitute slavery through biochemical means.[91] And although it is presented to readers as a possible future, what makes this scenario even more threatening is the fact that many of its elements already exist in the present.

Both the slave collar and hyperempathy are consistent with what Spillers calls "pornotroping," or the process by which black bodies are violently reduced to flesh. Just as the slave collar suspends the body in favor of the flesh, hyperempathy bypasses the constraints of the body to leverage the vulnerability of the flesh. What's more, pornotroping is a first-order process of racialization through which "race is constituted by a repeated sadistic white pleasure in black female suffering."[92] During her enslavement, Lauren discovers that her hyperempathy subjects her to both the pain of her fellow captives and the pornotropic pleasure of her captors; she writes: "there are a few men…who lash until they have orgasms. Our screams and convulsions and pleas and sobs are what these men need to feel sexually satisfied. I know of three who seem to need to lash someone to get sexual pleasure. Most often, they lash a woman, then rape her."[93] Once the Camp Christian "teachers" identify which of their captives were hyperempaths, "they were raped more often than the other women were."[94] In these scenes, the relationships between consent, pleasure, and body knowledge collapse as Lauren is forced to reenact the crisis of will and desire that characterizes the slave's existence.[95] While all hyperempaths would experience a similar crisis under these conditions regardless of race, black women are assured neither a restoration of their will and desire nor a discernible "end" to the crisis itself. This is to say that for Lauren, hyperempathy and the slave collar operate by intensifying an existing condition rather than imposing a new one. This structural vulnerability or "zero degree of social conceptualization" renders black female flesh as the site of production for modernity's symbolic order, where the value and meaning of our conceptual categories are both challenged and renewed.[96] Lauren performs an analogous function in the novels, as the knowledge gained from her hyperempathic episodes guides Earthseed's development into a global movement. In this context, Lauren's black life, or more precisely the blackness of her life, matters but only in its capacity to make all lives matter.

The approach to life promoted under the banner of Earthseed responds to our desires for new modes of existence appropriate for the Anthropocene. For Lauren, embracing change enables notions of self and community capable of adapting to complex socioeconomic, political, and environmental circumstances. In Earthseed, "god is a process or a combination of processes, not an entity. It is not conscious at all...God can be directed, focused, speeded, slowed, shaped. All things change, but all things need not change in all ways."[97] Moreover, change is not driven simply or only by historical progress. Each chapter in *Sower* and *Talents* opens with epigraphs from Earthseed's doctrinal text, *The Book of the Living*. Modeled after the aphoristic style of the Tao, the epigraph acknowledges the potential of political, economic, and social structures to affect and be affected by all matter: "We have lived before/We will live again/We will be silk,/Stone,/Mind,/Star,/We will be scattered,/ Gathered,/Molded,/Probed./We will live,/And we will serve Life."[98] The confluence of silk, stone, mind, and star rejects the idea that the active properties of "life" are confined to the human or organic, constituting what Weheliye calls a "radically different political imaginary," where "suffering appears as utopian erudition" that "[summons] forms of human emancipation that can be imagined but not (yet) described."[99] The destiny of Earthseed to "take root amongst the stars" is precisely this imagined yet indescribable emancipation.[100] Once the starships depart at the end of *Talents*—an exodus made possible by the recuperation of suspended animation and the reclamation of the hold—humanity becomes the seed of Earth, open to futures that we cannot predict as we spread to worlds unknown.

Visions like these suggest that oppressive conditions do not exhaust the variabilities of life, and that the transvaluation of matter and the human can encourage more comprehensive ethical orientations. Then again, perceiving hyperempathy and Earthseed as a means to "liberate… assemblages of life, thought, and politics from the tradition of the oppressed" obliges us to detach pornotroping from the sexually violent production of racial difference.[101] As a result, pornotroping is reframed as a vital disruption of the social and symbolic, one that produces a material body unbound by race. The body celebrated by literary critics and the new materialisms thus betrays a desire to harness the radical potential of black flesh without paying the social and historical costs of being black. Only then can an attunement to the material forces of difference be depicted as an ethical project with universal import. Certainly, in black women's "absence from a subject position," Spillers does locate the potential for a *sui generis* naming that claims the "insurgent ground" outside of "dominant symbolic activity."[102] The difficulty here is that the "monstrosity of a female with the potential to 'name'" emerges out of the histories of modern racial slavery and the transatlantic slave trade.[103] This is not to say that a capacity for ethical life does not exist under conditions of oppression, or that pornotroping is a totality from which nothing escapes. However, in order to confront effectively the causes and consequences of the Anthropocene, we first need to reckon with our social and libidinal investments in black flesh. This would require us to address how the entanglements of blackness, the nonhuman, and the human make only certain bodies and environmental configurations both legible and desirable.

By marshaling counter-memories of transatlantic slavery to establish counter-futures, Afrofuturism grants us a crucial viewpoint from which to survey the whole of the Anthropocene and beyond. This same viewpoint also makes clear the imbrication of technological development (i.e., space and seafaring) and slavery's disciplinary strategies (i.e., the ecology of the slave ship) in ways that are vital to revising our approaches to the Anthropocene. If we continue to allow the Anthropocene's philosophy of history and nature to determine our engagements with the environment, then black people and racial blackness will continue to inhabit the "position of the unthought."[104] To be clear, my objective is not to reject wholesale the new materialisms or scholarship on the nonhuman turn. Because a planetary-level crisis requires a planetary-level philosophy, their attempts to offer a broader theorization of matter and being are appropriate for our geological era. What I am suggesting instead is that challenges to human exceptionalism should proceed through a critique of race or we risk reorganizing old privileges ("All Lives") under new standards of being ("Matter").

Notes

1. Jared Sexton, "Unbearable Blackness," *Cultural Critique 90* (2015): 168.
2. Colin N. Waters, Jan Zalasiewicz, Colin Summerhayes et al., "The Anthropocene is Functionally and Stratigraphically Distinct from the Holocene," *Science* 351, no. 6269 (2016): 137–147.
3. Daniel Hartley, "Against the Anthropocene," *Salvage Zone* (2015).
4. Waters et. al., "Anthropocene," 139.
5. Nicholas Mirzoeff, "It's Not the Anthropocene, It's the White Supremacy Scene; or, The Geological Color Line," in *After Extinction*, ed. Richard Grusin (University of Minnesota Press, 2018): 140.
6. Diana Coole and Samantha Frost, "Introducing the New Materialisms," in *New Materialisms: Ontology, Agency, and Politics*, eds. Diana Coole and Samantha Frost (Duke University Press, 2010), 4.
7. Zakiyyah Iman Jackson, "Outer Worlds: The Persistence of Race in Movement 'Beyond the Human,'" *GLQ: A Journal of Lesbian and Gay Studies* 21, no. 2–3 (2015): 216.
8. Mirzoeff, "Not the Anthropocene," 124.
9. Alexander G. Weheliye, *Habeas Viscus: Racializing Assemblages, Biopolitics, and Black Feminist Theories of the Human* (Duke University Press, 2014), 21.
10. Many feminist, postcolonial, and critical race studies scholars insist that such studies must occur through an interrogation of race. See Rachel Lee, *The Exquisite Corpse of Asian America: Biopolitics, Biosociality, and Posthuman Ecologies* (New York University Press, 2014); and Mel Y. Chen,. *Intoxicated: Race, Disability, and Chemical Intimacy Across Empire* (Duke University Press, 2023).
11. Coole and Frost, "Introducing," 4 (emphasis in the original).

12. Giles Deleuze, *Difference and Repetition*, translated by P. Patton (Columbia University Press, 1968), 206.
13. Pheng Cheah, "Non-Dialectical Materialism," in *New Materialisms: Ontology, Agency, and Politics*, eds. Diana Coole and Samantha Frost (Duke University Press, 2010), 85–86.
14. Coole and Frost, "Introducing," 6.
15. Elizabeth Grosz, *The Nick of Time: Politics, Evolution, and the Untimely* (Duke University Press, 2004), 6.
16. Elizabeth Wilson, *Gut Feminism* (Duke University Press, 2015), 3.
17. Coole and Frost, "Introducing," 6.
18. Melissa A. Orlie, "Impersonal Matter," in *New Materialisms: Ontology, Agency, and Politics*, eds. by Diana Coole and Samantha Frost (Duke University Press, 2010), 118; Cheah, "Non-Dialectical," 89; Grosz, *Nick of Time*, 149; Jane Bennet, *Vibrant Matter: A Political Ecology of Things* (Duke University Press, 2010), 14.
19. Mei Zhan, "Cosmic Experiments: Remaking Materialism and Daoist Ethic Outside of the Establishment," *Medical Anthropology* 35, no. 3 (2016): 26.
20. Rosi Braidotti, "The Politics of 'Life Itself' and New Ways of Dying," in *New Materialisms: Ontology, Agency, and Politics*, eds. Diana Coole and Samantha Frost (Duke University Press, 2010), 214.
21. Stephen Tumino, "The Affective Turn in Pedagogy: The Ecstatic Teacher and Other Stories," *Rethinking Marxism* 23, no. 4 (2011): 555.
22. Karen Barad, *Meeting the Universe Halfway: Quantum Physics and the Entanglement of Matter and Meaning* (Duke University Press, 2007), 394; Jane Bennett, "Vibrant Matters: An Interview with Jane Bennett," Interviewed by Peter Gratton, *Philosophy in a Time of Error*, 2010; Orlie, "Impersonal Matter," 127; and William Connolly, "Materialities of Experience," in *New Materialisms: Ontology, Agency, and Politics*, eds. Diana Coole and Samantha Frost (Duke University Press, 2010), 196–197.
23. Braidotti, "Politics of 'Life Itself," 213.
24. Braidotti, "Politics of 'Life Itself," 214.
25. Connolly, "Materialities," 190–191.

26. Coole and Frost, "Introducing." Coole and Frost cite Althusser, Foucault, and some strains of neo-Marxism and ethnography as examples of similar approaches. Intriguingly, Ian Buchanan points out that these approaches can also yield observations so "obvious" that "one does not even need a concept to make this claim. This is history in the mode of one damn thing after another." Ian Buchanan, "Assemblage Theory and its Discontents," *Deleuze Studies* 9, no. 3 (2015): 388.
27. Bennett, "Vibrant Matters." To her credit, in an interview with Peter Gratton for his blog, "Philosophy in a Time of Error," Jane Bennett admits that she needs to "focus more carefully" on how assemblages assume the characteristics of repetition, duration, and stability; she writes: "I want to get better at discerning the topography of Becoming, better at theorizing the 'structural' quality of agentic assemblages. For the question of 'structure' — or maybe that is the wrong word, and the phrase you suggest below is better, i.e.„ 'linkages' between and within 'open relations' — does seem to fall in the shadow of the alluring image of an ever-free becoming…Inside a process of unending change, bodies and forces with duration are somehow emitted or excreted. But how?" While this is a positive development, I maintain that our scholarly activity is intimately shaped by the legacies of transatlantic slavery. Going forward then, the new materialisms must consider how blackness informs their major concepts or they risk reproducing the kinds of race-thinking that hold these legacies in place.
28. Hortense Spillers, "Mama's Baby, Papa's Maybe: An American Grammar Book," *Diacritics* 17, no. 2 (1987): 68.
29. Timothy Morton, *Ecology Without Nature: Rethinking Environmental Aesthetics* (Harvard University Press, 2007), 37.
30. In mapping black studies' tireless examination of the human and its others, Alexander Weheliye observes that there exists an equally long tradition "in which theoretical formulations by white European thinkers are granted a conceptual carte blanche, while those uttered from the purview of minority discourse that speak to the same questions are almost exclusively relegated to the jurisdiction of ethnographic locality." Weheliye, *Habeas Viscus*, 33.
31. Coole and Frost, "Introducing," 19.

32. Iris van der Tuin and Rick Dolphijn, *New Materialism: Interviews & Cartographies* (Open Humanities press, 2012), 88; 143.
33. Sara Ahmed, "Imaginary Prohibitions: Some Preliminary Remarks on the Founding Gestures of the 'New Materialism,'" *European Journal of Women's Studies*, 15, no.1 (2008): 23–39. Sara Ahmed argues that a politics of attention have dictated the new materialisms' "founding gestures" by allowing for a reading of feminist scholarship as fundamentally anti-biology. Her position paper provoked responses from Noela Davis and Iris van der Tuin the latter of whom comments more extensively on the debate in her co-edited collection of interview and essays, *New Materialism: Interviews and Cartographies*. What Ahmed's paper tracks is a pattern of disappointment with feminism's purported anti-biologism that stretches back to the early 1990s. As she argues, however, driving this disappointment is an assumption that feminist scholars should "know better"; particularly because a set of theories historically concerned with the body can and should be at the forefront of materialist innovations. While this does not seem like a ringing endorsement, the fact that new materialists claim (white) feminism as their generative field suggests that these feminisms have access to the "theoretical resources" to re-conceptualize the role of matter in embodiment. Or, at least those feminisms in which "material dimensions" have not yet "been marginalized by fashionable constructivist approaches and identity politics." Read with Coole and Frost's critique of "identity politics" as theoretically impoverished, it is clear that feminism is granted an intellectual complexity that "identity politics" are not. Coole and Frost, "Introducing," 19.
34. Dionne Brand, *A Map to the Door of No Return: Notes on Belonging* (Vintage Canada, 2001), 29.
35. Jackson, "Outer Worlds," 216; Brand, *Map to the Door*, 29.
36. Spillers, "Mama's Baby," 68.
37. Tapji Garba and Sara-Maria Sorentino, "Slavery is a Metaphor: A Critical Commentary on Eve Tuck and K. Wayne Yang's 'Decolonization is Not a Metaphor,'" *Antipode* 52, no. 3 (2020): 764–782.
38. Hortense Spillers, *Black, White, and In Color: Essays on American Literature and Culture* (University of Chicago Press, 2003), 380.

39. Octavia E. Butler, *Parable of the Talents* (Grand Central Publishing, 1998), 8.
40. Madhu Dubey, "Speculative Fictions of Slavery," *American Literature* 82, no. 4 (2010): 784.
41. Dubey, "Speculative Fictions," 799.
42. Naomi Klein, *The Shock Doctrine: The Rise of Disaster Capitalism* (Macmillan, 2007).
43. Octavia E. Butler, *Parable of the Sower* (Warner Books, 1993): 120–121.
44. Butler, *Sower* 121; 118.
45. Butler, *Sower*, 118.
46. Butler, *Sower*, 292.
47. Butler, *Sower*, 176.
48. Peter G. Stillman, "Dystopian Critiques, Utopian Possibilities, and Human Purposes in Octavia Butler's Parables," *Utopian Studies* 14, no. 1 (2003): 23.
49. Butler, *Sower*, 287.
50. Butler, *Sower*, 288.
51. Butler, *Sower*, 288.
52. Patricia Melzer, "'All That You Touch You Change': Utopian Desire and the Concept of Change in Octavia Butler's *Parable of the Sower* and *Parable of the Talents*," *FemSpec* 3, no. 2 (2002): 41.
53. In their introduction to the *Science Fiction Studies* tribute to Octavia Butler, editors De Witt Kilgore and Ranu Samantrai write that "Butler approached sf askance, choosing to write self-consciously as an African American woman marked by a particular history. Her example clarifies the stakes for any particular minority breaking into forms seen as ethnically exclusive: the necessity or simply the desire to see oneself complexly represented in one's culture. Butler entered the field at a time when science fiction did not serve that function for white women or for.

    people of color." De Witt Douglas Kilgore and Ranu Samantrai, "A Memorial to Octavia E. Butler," *Science Fiction Studies* 37, no. 3 (2010): 353.
54. Octavia Butler, Marilyn Mehaffy, and AnaLouise Keating, "'Radio Imagination': Octavia Butler on the Poetics of Narrative Embodiment," *Melus* 26, no. 1 (2001): 46.

55. Gregory J. Hampton, *Changing Bodies in the Fiction of Octavia Butler: Slaves, Aliens, and Vampires* (Lexington Books, 2010), 69.
56. Butler et al., "Radio Imagination," 59.
57. Butler, *Sower*, 11.
58. Butler, *Sower*, 12.
59. Butler, *Sower*, 278.
60. Jim Miller, "Post-Apocalyptic Hoping: Octavia Butler's Dystopian/Utopian Vision," *Science Fiction Studies* 25, no. 2 (1998): 357; Stillman, "Dystopian Critiques," 29; Melzer, "Touch," 13.
61. Phillip E. Wegner, "Jameson's Modernisms; or, the Desire Called Utopia," *Diacritics* 37, no. 4 (2007): 17.
62. Cole and Frost, "Introducing," 10.
63. Butler, *Sower*, 297.
64. Diana Coole, "The Inertia of Matter and the Generativity of Flesh," in *New Materialisms: Ontology, Agency, and Politics,* eds. Diana Coole and Samantha Frost (Duke University Press, 2010), 104.
65. Jerry Phillips, "The Intuition of the Future: Utopia and Catastrophe in Octavia Butler's Parable of the Sower," *Novel: A Forum on Fiction* 35, no. 2–3 (2002): 306.
66. Butler, *Sower*, 45–46.
67. Braidotti, "Politics of 'Life Itself,'" 210.
68. Butler, *Sower*, 196.
69. Butler, *Sower*, 36.
70. Butler, *Sower*, 208.
71. Melzer, "Touch," 12; Phillips, "Intuition," 307.
72. Butler, *Sower*, 115.
73. Octavia Butler, "On Racism," NPR. Last modified August 20, 2001. https://www.npr.org/2001/08/20/5245679/on-racism.
74. Butler, *Sower*, 305.
75. For an example of a similar reimagining, see Rivers Solomon, *An Unkindness of Ghosts* (Akashic Books, 2017).
76. Butler, *Sower*, 261.
77. Butler, *Sower,* 261; *Talents* 179.
78. Bulter, *Talents*, 406.
79. Butler, *Talents*, 407.
80. Butler, *Talents*, 406.

81. Spillers, "Mama's Baby," 67.
82. Spillers, "Mama's Baby," 67.
83. Butler, *Talents*, 84.
84. Butler, *Talents*, 131.
85. Zakiyyah Iman Jackson, *Becoming Human: Matter and Meaning in an Antiblack World* (New York University Press, 2020), 161–162; Sylvia Wynter, "Towards the Sociogenic Principle: Fanon, Identity, the Puzzle of Conscious.

    Experience, and What It Is Like to be 'Black,'" in *National Identities and Sociopolitical Changes in Latin America*, eds. Mercedes F. Durán-Cogan and Antonio Gómez Moriana (Routledge, 2013).
86. Butler, *Sower*, 292–293.
87. Spillers, "Mama's Baby," 67.
88. Butler, *Talents*, 241.
89. Butler, *Talents*, 239.
90. Butler, *Talents*, 227–228.
91. Butler explores this theme more fully in the first novel of her *Lilith's Brood* trilogy. See Octavia Butler, *Dawn* (Grand Central Publishing, 1987).
92. Jennifer Nash, *The Black Body in Ecstasy: Reading Race, Reading Pornography* (Duke University Press, 2014), 52.
93. Butler, *Talents*, 233.
94. Butler, *Talents*, 242.
95. See Saidiya Hartman's *Scenes of Subjection*, especially chapter 3, "Seduction and the Ruses of Power."
96. Spillers, "Mama's Baby," 67.
97. Butler, *Talents*, 46.
98. Butler, *Talents*, 60.
99. Weheliye, *Habeas Viscus*, 126–127.
100. Butler, *Talents*, 46.
101. Weheliye, *Habeas Viscus*, 137.
102. Spillers, "Mama's Baby," 68.
103. Spillers, "Mama's Baby," 80.
104. Saidiya Hartman and Frank B. Wilderson III, "The Position of the Unthought," *Qui Parle* 13, no. 2 (Spring/Summer 2003): 185; 188.

References

Ahmed, Sara. "Imaginary prohibitions: Some Preliminary Remarks on the Founding Gestures of the 'New Materialism.'" *European Journal of Women's Studies*, 15, no.1 (2008): 23-39.

Barad, Karen. *Meeting the Universe Halfway: Quantum Physics and the Entanglement of Matter and Meaning*. Duke University Press, 2007.

Bennett, Jane. *Vibrant Matter: A Political Ecology of Things*. Duke University Press, 2010.

Bennett, Jane. "Vibrant Matters: An interview with Jane Bennett." Interviewed by Peter Gratton. *Philosophy in a Time of Error.* 2010. Retrieved from: http://philosophyinatimeoferror.com/2010/04/22/vibrant-matters-an-interview-with-jane-bennett/

Braidotti, Rosi. "The Politics of 'Life Itself' and New Ways of Dying." In *New Materialisms: Ontology, Agency, and Politics*, edited by Diana Coole and Samantha Frost. Duke University Press, 2010.

Brand, Dionne. *A Map to the Door of No Return: Notes on Belonging*. Vintage Canada, 2001.

Buchanan, Ian. "Assemblage Theory and its Discontents." *Deleuze Studies* 9, no. 3 (2015): 382-392.

Butler, Octavia E. *Dawn*. Grand Central Publishing, 1987.

Butler, Octavia E. "On Racism." NPR. Last modified August 20, 2001. https://www.npr.org/2001/08/20/5245679/on-racism

Butler, Octavia E. *Parable of the Sower*. Warner Books, 1993.

Butler, Octavia E. *Parable of the Talents*. Grand Central Publishing, 1998.

Butler, Octavia E., Marilyn Mehaffy, and AnaLouise Keating. "'Radio Imagination': Octavia Butler on the Poetics of Narrative Embodiment." *Melus* 26, no. 1 (2001): 45-76.

Callaway, Elizabeth. *Eden's Endemics: Narratives of Biodiversity on Earth and Beyond*. University of Virginia Press, 2020.

Cheah, Pheng. "Non-Dialectical Materialism." In *New Materialisms: Ontology, Agency, and Politics*, edited by Diana Coole and Samantha Frost. Duke University Press, 2010.

Chen, Mel Y. *Intoxicated: Race, Disability, and Chemical Intimacy Across Empire*. Duke University Press, 2023.

Connolly, William. "Materialities of Experience." In *New Materialisms: Ontology, Agency, and Politics*, edited by Diana Coole and Samantha Frost. Duke University Press, 2010.

Coole, Diana. "The Inertia of Matter and the Generativity of Flesh." In *New Materialisms: Ontology, Agency, and Politics*, edited by Diana Coole and Samantha Frost. Duke University Press, 2010.

Coole, Diana and Samantha Frost. "Introducing the New Materialisms." In *New Materialisms: Ontology, Agency, and Politics*, edited by Diana Coole and Samantha Frost. Duke University Press, 2010.

Deleuze, Giles. *Difference and Repetition*, Translated by P. Patton. Columbia University Press, 1968.

Dubey, Madhu. "Speculative Fictions of Slavery." *American Literature* 82, no. 4 (2010): 779-805.

Garba, Tapji and Sara-Maria Sorentino. "Slavery is a Metaphor: A Critical Commentary on Eve Tuck and K. Wayne Yang's 'Decolonization is Not a Metaphor." *Antipode* 52, no. 3 (2020): 764-782.

Grosz, Elizabeth. *The Nick of Time: Politics, Evolution, and the Untimely*. Duke University Press, 2004.

Grosz, Elizabeth. "Feminism, Materialism, and Freedom." In *New Materialisms: Ontology, Agency, and Politics*, edited by Diana Coole and Samantha Frost. Duke University Press, 2010.

Grusin, Richard. *The Nonhuman Turn*. University of Minnesota Press, 2015.

Hampton, Gregory J. *Changing Bodies in the Fiction of Octavia Butler: Slaves, Aliens, and Vampires*. Lexington Books, 2010.

Hartley, Daniel. "Against the Anthropocene." *Salvage Zone* (2015).

Hartman, Saidiya. *Scenes of Subjection: Terror, Slavery, and Self-Making in Nineteenth Century America*. Oxford University Press, 1997.

Hartman, Saidiya and Frank B. Wilderson III. "The Position of the Unthought." *Qui Parle* 13, no. 2 (Spring/Summer 2003): 183–201.

Heise, Ursula K. *Imagining Extinction: The Cultural Meanings of Endangered Species*. University of Chicago Press, 2019.

Jackson, Zakiyyah Iman. *Becoming Human: Matter and Meaning in an Antiblack World*. New York University Press, 2020.

Jackson, Zakiyyah Iman. "Outer Worlds: The Persistence of Race in Movement 'Beyond the Human.'" *GLQ: A Journal of Lesbian and Gay Studies* 21, no. 2.3 (2015): 215–218.

Kilgore, De Witt Douglas, and Ranu Samantrai. "A Memorial to Octavia E. Butler." *Science Fiction Studies* 37, no. 3 (2010): 353–361.

Klein, Naomi. *The Shock Doctrine: The Rise of Disaster Capitalism*. Macmillan, 2007.

Lee, Rachel. *The Exquisite Corpse of Asian America: Biopolitics, Biosociality, and Posthuman Ecologies*. New York University Press, 2014.

Liboiron, Max. *Pollution is Colonialism*. Duke University Press, 2021.

Melzer, Patricia. "'All That You Touch You Change': Utopian Desire and the Concept of Change in Octavia Butler's *Parable of the Sower* and *Parable of the Talents*." *FemSpec* 3, no. 2 (2002): 31-52.

Miller, Jim. "Post-Apocalyptic Hoping: Octavia Butler's Dystopian/Utopian Vision." *Science Fiction Studies* 25, no. 2 (1998): 336-360.

Mirzoeff, Nicholas. "It's Not the Anthropocene, It's the White Supremacy Scene; or, The Geological Color Line." In *After Extinction*, edited by Richard Grusin. University of Minnesota Press, 2018.

Morton, Timothy. *Ecology Without Nature: Rethinking Environmental Aesthetics.* Harvard University Press, 2007.

Nash, Jennifer. *The Black Body in Ecstasy: Reading Race, Reading Pornography.* Duke University Press, 2014.

Orlie, Melissa A. "Impersonal Matter." In *New Materialisms: Ontology, Agency, and Politics*, edited by Diana Coole and Samantha Frost. Duke University Press, 2010.

Phillips, Jerry. "The Intuition of the Future: Utopia and Catastrophe in Octavia Butler's *Parable of the Sower*." *Novel: A Forum on Fiction* 35, no. 2–3 (2002): 299–311.

Sexton, Jared. "Unbearable Blackness." *Cultural Critique 90* (2015): 159-178.

Solomon, Rivers. *An Unkindness of Ghosts*. Akashic Books, 2017.

Spillers, Hortense. *Black, White, and In Color: Essays on American Literature and Culture.* University of Chicago Press, 2003.

Spillers, Hortense. "Mama's Baby, Papa's Maybe: An American Grammar Book." *Diacritics* 17, no. 2 (1987): 65-81.

Stillman, Peter G. "Dystopian Critiques, Utopian Possibilities, and Human Purposes in Octavia Butler's Parables." *Utopian Studies* 14, no. 1 (2003): 15-35.

Tsing, A.L., Deger, J., Saxena, A.K. and Zhou, F. *Field Guide to the Patchy Anthropocene: The New Nature.* Stanford University Press, 2024.

Tuin, Iris van der, and Rick Dolphijn. *New Materialism: Interviews & Cartographies.* Open Humanities press, 2012.

Tumino, Stephen. "The Affective Turn in Pedagogy: The Ecstatic Teacher and Other Stories." *Rethinking Marxism* 23, no. 4 (2011): 548-556.

Waters, Colin N., Jan Zalasiewicz, Colin Summerhayes et al. "The Anthropocene is Functionally and Stratigraphically Distinct from the Holocene." *Science* 351, no. 6269 (2016): 137–147.

Weheliye, Alexander G. *Habeas Viscus: Racializing Assemblages, Biopolitics, and Black Feminist Theories of the Human.* Duke University Press, 2014.

Wegner, Phillip E. "Jameson's Modernisms; or, the Desire Called Utopia." *Diacritics* 37, no. 4 (2007): 2-20.

Wilson, Elizabeth A. *Gut Feminism.* Duke University Press, 2015.

Wynter, Sylvia. "Towards the Sociogenic Principle: Fanon, Identity, the Puzzle of Conscious Experience, and What It Is Like to be 'Black.'" In *National Identities and Sociopolitical Changes in Latin America.* Edited by Mercedes F. Durán-Cogan and Antonio Gómez Moriana. Routledge, 2013.

Zhan, Mei. "Cosmic Experiments: Remaking Materialism and Daoist Ethic Outside of the Establishment." *Medical Anthropology* 35, no. 3 (2016): 247-262.

Epilogue

Oceanic Feeling: (Anti)Blackness in **The Deep**

In July 2020, the words "racist fish" were spray painted across the base of Copenhagen's famed Little Mermaid statue. The incident was covered widely by the international press and due to its timing, news outlets were quick to connect it to the Black Lives Matter movement and the global protests it inspired. The inscription appeared to implicate Hans Christian Andersen's fairytale in the same antiblack systems against which the protests were arrayed. On the one hand, the statue was not a surprising choice for anti-racist messaging. As one of the most photographed and visited attractions in the world, it has become a popular site for political vandalism, with activists calling attention to a variety of causes through its defacement or disfiguration. The words "Free Hong Kong" were scrawled on its base in January of that same year. On the other hand, the inscription offered little clue as to how or why the mermaid is related to race. French broadcast network *France 24* acknowledged this lack of clarity, observing that "the apparent racist accusation" had left police and spectators "scratching their heads."[1] In an attempt to explain, several articles cited Disney's live-action remake of *The Little Mermaid* and the "controversy" over the casting of black actress Halle Bailey in the title role.[2]

When Disney began promoting the film in 2019, the announcement of her casting was met with a wave of racist abuse, most of which

D. Leong, *Black Literature, Social Justice, and the Environment*,
https://doi.org/10.1007/978-3-032-19288-2

was directed at Bailey herself. As Jessica Pressman points out, "white supremacists may have had a hard time coming to terms with the fact that a woman of color, not a white woman, figures as a character that viewers of any color can identify with and adore." This is to say that the backlash to the film was less about the "mythological figure of the mermaid" and more about preserving the "myth of whiteness as a cultural standard."[3] While these details would have clarified the meaning of the accusation, very few articles went beyond their initial reference to the film and its controversy. They relied instead on a quotation from Andersen scholar Ane Grum-Schwensen who commented, "I am having a hard time seeing what is particularly racist in the fairytale *The Little Mermaid*."[4] By choosing not to elaborate on the backlash to the film, the articles further mystified the accusation by framing it as a misreading of Andersen's story. As a result, they precluded more careful consideration of how the mermaid—as a "stake, symbol, and unstable signifier in debates about cultural ownership and identity"—is entangled with race.[5] The reporting on this incident was more or less consistent with how the popular press covered the Black Lives Matter movement and the broader issues it raised. It is notable, therefore, that it was presented in isolation from the murder that fueled nationwide debates over antiblackness in Denmark.

On June 23, 2020, black Danish citizen Phillip Mbuji Johansen was tortured and murdered by brothers Magnus and Mads Moeller, both of whom are white. Over the course of a night, the brothers broke his skull and legs, crushed his testicles, branded his face and torso with a hot iron, beat him with a wooden pole, and stabbed him 39 times. The attack bore many of the hallmarks of a Jim Crow-Era lynching: brutality, mutilation, intimate violence, and even an accusation of sexual assault. The brothers argued in court that "they had not intended to kill their victim" and only wanted "to teach him a lesson because they believed he had made sexual advances to their mother."[6] While the case drew significant attention from the outset, it became a matter of national and international interest after prosecutors declined to charge the brothers with a hate crime. This was despite the fact that Mads Moller had swastika and white power tattoos and had expressed public support for White Lives Matter. Police and prosecutors concluded that because Johansen knew his attackers, "the killing had nothing to do with 'skin color'" and was instead "a personal relationship…gone wrong."[7] In their view, if "there was a personal motive in the killing, it could not also be a hate crime."[8] This rationale was criticized strongly by Danish activists, scholars, and progressive politicians, many of whom called for investigations into structural racism and judicial bias.[9]

Because Johansen was killed only a month after George Floyd, virtually every aspect of the case from the initial investigation to the Mollers' sentencing was discussed in relation to Floyd and the Black Lives Matter movement. For example, Danish news outlets reported that "a knee was pressed hard against Mr. Johansen's neck, just as in the killing of Mr. Floyd," while prosecutor Benthe Pedersen Lund told newspaper *Bornholms Tidende* that her office was "fully aware of the situation in the USA."[10] It was precisely during this period and in this racial climate that the Little Mermaid statue was vandalized. And yet, even with a clear context for the vandalism if not the message itself, only the *Guardian* and *Washington Post* briefly link it to Johansen's murder while none of the reporting on Johansen's murder refers to the vandalism.[11] The *Guardian* even notes that the statue was tagged the day after "activists accused Danish officials of being unable to recognise racism after authorities said the killing of a biracial man by two white men was not racially motivated."[12] The charge of being "unable to recognize racism" was certainly borne out by reactions to the phrase "racist fish" though perhaps not in the way activists imagined it would.

There are several factors that might account for the separation of the two incidents. Like Hans Christian Andersen himself, the mermaid installation is a national icon and the accusation of racism arrived when residents were attempting to distance Danish racial politics from those of the United States.[13] In addition, reporting on the Black Lives Matter movement often implied that anti-racist actions were confusing or evidence of hysterical overreach, including the defacing of statues. And while the mermaid's multicultural history is now a subject of popular and scholarly discourse, the mermaid itself is not associated primarily with race.[14] At the same time, the separation of these events is too uniform to be accidental or incidental, especially since both events were covered almost exclusively in relation to the Black Lives Matter movement. Whether deliberate or unintentional, the separation had the same effect as Disney's casting controversy and the reporting on the accusation itself. In each case, a demand for black liberation was minimized or mystified in favor of preserving whatever racial myths the mermaid represents. Accordingly, the mermaid not only indexes our relationships to and concepts of racial difference but also offers new ways of understanding the ecological legacies of slavery and the slave ship.

This epilogue explores how white racial fantasies and black liberatory desire are negotiated through the figure of the mermaid and the oceanic imaginaries it represents. By taking the incidents above as a point of departure, it aims to show how the psychic life of race is entangled with the history and materiality of the ocean. This effort builds on recent work conducted at the intersections of the blue humanities and black studies. It has become essential to develop new ways of thinking with and through water and to this end the blue humanities draw from literary and cultural studies, the social sciences, and the creative arts to probe how humans and nonhumans shape and are shaped by aquatic concerns.[15] By engaging water's fluid representations, the practices that give it meaning, and its frameworks of theorization, the blue humanities seek to clarify how water functions as myth, metaphor, material, and medium. Similarly, scholars of race and the environment have examined how the Middle Passage serves as a formative site for the Anthropocene and our concepts of nature and race.[16] To confront the "ever-present specter of the transatlantic slave trade," Joshua Bennett charts a black hydropoetics that operates as "the intersection of terra firma and open sea, surface and benthos, [and] the observable ocean and the uncharted blackness of its very bottom."[17] Alexis Pauline Gumb's *Undrowned: Black Feminist Lessons from Marine Mammals* explores the possibilities of a "trans-species communion" based on the "subversive and transformative guidance of marine mammals."[18] And Jonathan Howard's blue and black humanities propose the "the middle passages of those [drowned Africans] remembered by Equiano as the '*inhabitants* of the deep'" as a powerful site from which to rethink the human.[19]

As this scholarship suggests, when we attend to the legacies of slavery and the slave trade, the Middle Passage and the ocean more broadly emerge as spaces of "otherworldly despair and fugitive possibility," ones in which an "uncharted, undercommon marronage [is] made possible by the opacity of the oceanic realm."[20] In one of the most powerful commentaries on the Middle Passage, Hortense Spillers observes with respect to this despair and possibility that:

> Those African persons in "Middle Passage" were literally suspended in the "oceanic," if we think of the latter in its Freudian orientation as an analogy for undifferentiated identity: removed from the indigenous land and culture, and not-yet "American" either, these captive persons, without names that their captors would recognize, were in movement across the

> Atlantic, but they were also *nowhere* at all. Inasmuch as, on any given day, we might imagine, the captive personality did not know where s/he was, we could say that they were the culturally "unmade"...Often enough for the captains of these galleys, navigational science of the day was not sufficient to guarantee the intended destination. We might say that the slave ship, its crew, and its human-as-cargo stand for a wild and unclaimed richness of *possibility* that is not interrupted, not "counted"/ "accounted," or differentiated, until its movement gains the land thousands of miles away from the point of departure.[21]

In locating Freud's oceanic feeling in the Middle Passage, Spillers provides this mystical sensation with a material history and cultural specificity. Consequently, the "undifferentiated identity" for which we yearn is found not only in a fantasy of "eternal, unbounded, and interconnected plenitude" but also in the spacetime of the slave ship and its very real oceanic environment.[22] This latter point is reinforced by Spillers' reminder about the dangers of transatlantic sailing. As I detail in the previous chapters, deep sea vessels often strayed off course, ran out of supplies, battled outbreaks of disease, engaged in armed conflict, and negotiated hazardous and unpredictable weather. The slave ship environment was characterized by a shared precarity and as Spillers notes, until it reached the shore, sailors (i.e., "crew"), slaves, and the ship itself stood "for a wild and unclaimed richness of possibility." More importantly, while the hold's "undifferentiated identity" contributed to this richness, it remained specific to the slaves as an effect of their cultural unmaking. On a symbolic level, the slave's suspension in the "oceanic" introduced an anomaly into the ship's environment, namely a sensation or state of being that is desirable to sailors but available only to slaves. We might therefore consider the slave ship and its ecology as a model for understanding how race in general and (anti)blackness in particular animate the power and appeal of the oceanic.

To this end, I take my cue from Spillers and Rivers Solomon's Afrofuturist novel *The Deep* to sketch a speculative genealogy for oceanic feeling. Solomon's novel centers on a community of merfolk known as the wajinru, all of whom are descendants of pregnant slaves thrown overboard during the crossing of the Middle Passage. To avoid the weight of their traumatic history, all memories of the wajinru's origins are carried by a designated historian. As the narrative makes clear, the wajinru's methods of communication, their embeddedness in multispecies networks, and even the very matter and material of their memories are inseparable from the ocean itself. By reading Solomon's novel alongside the conceptual development of Freud's oceanic feeling, I argue that *The Deep* depicts oceanic feeling as a state of radical passivity in which difference is experienced as possibility itself. Moreover, the novel links this state fundamentally to black liberatory desire in a way that clarifies the relationship between the slave ship and the oceanic. The fear and fantasy that black people have privileged or exclusive access to freedom, power, or enjoyment have long shaped the politics of race. Jared Sexton has demonstrated, for instance, that the pro-life investment in black birth is based on a "hallucinatory projection of the primordial, elemental, undifferentiated vita that blacks, in the racist imagination, are supposed to enjoy." In this "fantasy of prenatal nonmorality," the black womb and fetus come to represent the "true preoccupation of the pro-life movement," or "the total cancellation of desire and, hence, the elimination of pleasure and pain, by an uninterrupted enjoyment."[23] It is not coincidental that oceanic feeling is conceived as an "original maternal plenitude" or the sensation of a womb-like space.[24]

When we bring the ecology of the slave ship to bear on the ocean and the oceanic, and on the mermaid and the fetus, the hostility to Bailey's casting appears as a kind of racial resentment. If the mermaid's fluid, liminal existence is central to her appeal, then beneath the attachment to her whiteness is a desire for a freedom or existence that black people are suspected already to enjoy. Whether figured as a sensation of "limitless being" or a state of pure possibility, the slave's suspension in the oceanic helps to drive its appeal.[25] To clarify, my objective in assembling these texts, events, and representations is not to critique or reconstruct the prevailing intellectual genealogy of oceanic feeling. My approach is modeled instead on what Saidiya Hartman calls critical fabulation, which involves "playing with and rearranging the basic elements of [a] story" and "re-presenting the sequence of events in divergent stories and from contested points of view."[26] By "throwing into crisis 'what happened when' and by exploiting the 'transparency of sources' as fictions of history," I hope to find creative ways to intervene in the ecology of the slave ship by making legible the entanglements on which it depends.[27]

Deep in the Unconscious

Before becoming the model of white, heteronormative femininity popularized by Disney's animated 1989 film, the mermaid was already a near-universal figure of folklore. Depending on their cultural and historical origins, these figures are portrayed variously as vengeful or benevolent, monstrous or sacred, and deviant or idealized. Mermaid lore also received a major boost during the Columbian Exchange and transatlantic slavery. During these periods, anxieties over novel forms of difference were mapped often onto its half-human and half-fish form. Perhaps the most famous example concerns P.T. Barnum's "Fiji Mermaid," a specimen that was exhibited in New York and Boston during the 1840s and was later revealed to be the head and torso of a monkey sewn to a fish. Notably, it was because of and not despite its questionable authenticity that the Fiji Mermaid attracted considerable public interest. By visibly transgressing the categories of gender, species, and race, Barnum's specimen "[bore] out emerging scientific narratives that understood racial others, gender deviants, and disabled individuals to be 'missing links' and evolutionary throwbacks."[28] In this respect, the mermaid is not only a symbol of hybridity and liminality but also stands in for the unresolved contradictions within our social orders. It is hardly surprising that its historical appearances tend to coincide with periods of heightened racial anxiety. Columbus's first mention of mermaids, for instance, comes from a journal entry recorded in Haiti in 1493.[29]

These racial currents shifted when electronic music duo Drexciya released their 1997 compilation album entitled *The Quest*. The album's liner notes, which are credited to an "unknown writer," speculate about a community of aquatic beings descended from pregnant African slaves thrown overboard in the Middle Passage. The writer asks: "Could it be possible for humans to breath underwater? A fetus in its mother's womb is certainly alive in an aquatic environment…Is it possible that they could have given birth at sea to babies that never needed air?"[30] In Drexciya's reimagining of the Black Atlantic, the sea is not only a material archive for slavery but also in a literal sense the birthplace of its afterlife. Or more precisely, its afterlives. The mythology surrounding these "water-breathing, aquatically mutated" beings expanded in 2019 with the release of two related texts—a song entitled "The Deep" by experimental hip-hop group clipping. and Solomon's novel.[31]

The Deep centers on a community of Drexciyan merfolk known as the wajinru. When the narrative opens, the wajinru are scattered across the ocean in sparsely populated settlements, their once thriving cities decimated by climate change and ocean mining. To avoid the trauma associated with their origins and history, their collective memories are carried by a designated "historian." The community gathers for a ritual of "Remembrance" once a year, which takes place in a "giant mud sphere" called "the womb."[32] During these rituals, the wajinru's memories are transferred temporarily from historian to community in and as a series of embodied experiences. But for Yetu, the current historian and novel's protagonist, the burden of channeling the memories of "every wajinru who had ever lived" has become unbearable.[33] Where "she had once imagined channeling as a sweet, beautiful flow of energy, the past running gently through her," it was now "more like slitting an opening in herself so they could get out."[34] As a consequence, she flees during a remembering and leaves the wajinru without a guide to navigate their reliving of history or the frenzied grief it unleashes. After swimming to a nearby landmass, she becomes stranded in a tidepool and soon enters into a queer, interspecies relationship with an African human or "two legs" named Oori. Yetu then returns to the wajinru who, in her absence, have embraced the power of a history that is living and shared. The novel closes with Yetu inviting Oori into sea to experience her memories, after which Oori develops the ability to breathe in water and on land. As this summary suggests, the wajinru's racial memories, history, communication, and kinship relations are fluid, diffuse, and intensely felt. Their way of being depends on the ocean's physical properties in such a way that the distinctions between self and other, subject and environment, and meaning and matter are mutable. The novel, in other words, renders oceanic feeling both literal and material.[35]

As a psychoanalytic concept, oceanic feeling developed out of the correspondence between Freud and French writer Romain Rolland, the latter of whom was a student of several mystic traditions. In 1927, Freud sent a copy of *The Future of an Illusion* to Rolland who expressed concern that Freud had not adequately addressed the source of religious experience. In asking him to take up again the question of religion, Rolland described a "religious feeling" that while separate from any organized doctrine, "was the simple and direct fact of the sensation of the eternal…without perceptible limits, and like the ocean."[36] In return Freud confessed he had no experience with such a feeling but responded nevertheless in *Civilization and Its Discontents* (hereafter *Civilization*). To do so, he traced Rolland's "sensation of the eternal" to the pre-Oedipal stage of psychosexual development. Within this stage, the primary narcissistic union between mother and infant is experienced as an "indissoluble bond of being one with the external world."[37] The sensation or feeling Rolland described was thus a "fantasmatic effect of the subject's subjection to language" and the concomitant loss or repression of the maternal.[38] As Claire Colebrook elaborates, for Freud "the oceanic is a structure that helps explain the affect of a necessarily tragic world. From a feeling of isolation, finitude, and privation one assumes that there *must have been* a lost origin of connectedness…It is an indication that the world is intolerable, that one yearns for an impossible outside."[39]

While this chapter of psychoanalytic history has been thoroughly examined, less familiar is the role that Haitian creole mesmerism played in the recognition of the unconscious and the stoking of the Haitian Revolution. Mesmerism, or "animal magnetism," arrived in Haiti in 1784 via the Comte de Chastenet Antoine-Hyacinth de Puységur who was himself a student of Anton Mesmer. For several years, the Comte and his brother the Marquis de Puységur conducted therapeutic sessions for slaves, colony officials, and plantation owners alike. So successful were these sessions that they began to threaten the authority of the colony's medical and scientific communities. As a result, the colony's first scientific society was established in part to disprove the science behind Mesmer's doctrine and thereby diminish its popularity. However, the practice had already become widespread among Haiti's enslaved population. Slaves from several plantations would gather for late night mesmerist rituals, or "night assemblies," during which weapons were made and exchanged, talismans were sold, and according to one plantation owner's account, "most participants asked for the ability to control the mind of their masters."[40] Unsurprisingly, the colony's response shifted away from scientific standardization and toward violent repression. As Nathan Gorelick explains, "when [this] wave of clandestine creole mesmerism swept across the colony late in the 1780s, it encouraged disobedience, desertion, self-determination, and rebellion, and, alongside or perhaps within Vodou, was an important component of the revolutionary collective consciousness that culminated with the Haitian Revolution."[41] Certainly, creole mesmerism is neither psychology nor psychoanalysis, but in individual sessions and the "late night assemblies," the magnetic rapport established between leaders and participants was in essence a precursor to what Freud would later describe as transference. As Gorelick argues, "mesmerism, regardless of where or among whom it was practiced, was a clinically rudimentary means of drawing out the unconscious from behind the veil of repression"; in this sense we might say that "creole mesmerism was an encounter with the unconscious, before it knew its own name."[42] While we cannot draw a straight line from these revolutionary encounters with the black unconscious to Freud's dismissal of mysticism, this speculative entanglement offers a new way to approach oceanic feeling in Solomon's novel.

The Deep establishes oceanic feeling as the effect of a therapeutic encounter with black liberatory desire, one that occurs specifically within a shared environment. Like the late night assemblies of creole mesmerism, the wajinru's rituals of Remembrance begin when "mind [leaves] body."[43] In this state, "the entirety of the wajinru people would be entranced by the History. They would move, but according to instinct and random pulses in their brains, indecipherable from a seizure."[44] Yetu, rather than "tell[ing]" the Remembrance," helps "her people experience it as it happened in the minds of various wajinru who lived it."[45] This is to say that the Remembrance makes literal and material the process of transference between the historian-as-analyst and the wajinru-as-analysands. The medium and method for this transference, moreover, is electromagnetic charge: "In addition to reaching into the minds of the wajinru to log the events of the era historians learned whose minds were electro-sensitive enough to host the rememberings in the future."[46] The Remembrance succeeds in large part because the wajinru cannot resist the historian's "magnetic energy."[47] Indeed, it seems to establish temporarily for Yetu the state of "oneness" or undifferentiated plenitude to which Spillers and Freud refer: "Yetu let herself feel her body. The cold, heavy water. The brush of currents. Her mind was empty. Blessedly empty. Nothingness filled her in a rush of heat. There was no one there. No voices, no stories, no visions of deaths. Blissed out, she floated limply until she could no longer ignore the pings of water coming from the other wajinru."[48] However, because this state is induced specifically by the historian's loss of history, it is instead a caricature or even a fantasy of what oceanic feeling could be: "[Yetu] imagined how it felt when the History left her, the freedom of it, but if freedom only brought loneliness, emptiness, what was the point? Nothingness was a fate worse than pain. How long would it take Yetu to become ravenous for something to fill the hole the way other wajinru did?"[49] It is only after the wajinru's history is redistributed across their community that the feeling and experience of the mud womb becomes properly oceanic. This point is reinforced during the conclusion of the novel when Yetu's mother shares: "There is very little difference between a bornt wajinru pup and one still encased in the womb. What if some of your rememberings of dark loneliness as a pup were you inside a belly, and it was hardly distinguishable from floating in the deep? It is all waters."[50] In this way, *The Deep* depicts oceanic feeling not as a desire for an "illusory maternal matrix" or a fantasy of uninterrupted enjoyment, but as the result of a sustained engagement with a black revolutionary unconscious.[51] We might even conceive of Solomon's novel and black literature more broadly as a means of working through our complex and often interdicted relationships to black freedom.

If we read Freud's correspondence with Rolland through this lens, his treatment of the oceanic acquires new significance. Although he agreed generally with Freud's critique of organized religion, Rolland was determined to show that mysticism was not simply "one more form of the 'common-man's religion.'"[52] To do so, he distinguished between religious sentiment as "*allegiance* to or *faith* in a dogma, or a word of god (or a tradition)" and religious sentiment as "free *vital upsurge*." Unlike the former, he claimed, the latter was a "constant state" that had "nothing to do with [his] personal yearnings" and did not affect his "critical faculties" or the "freedom to exercise them."[53] Whereas the oceanic for Freud was an escapist "yearn[ing] for an impossible outside," it was for Rolland a way of being or mode of existence. Oceanic feeling in his view was "dynamic, vitalistic, creative, socially adaptive" and ultimately "beneficial to mankind."[54] Indeed, in situating oceanic feeling within his model of development, Freud either misunderstood or ignored Rolland's suggestion that the oceanic is a lasting orientation to the world, one that could be achieved over time.[55] William Parsons notes apropos of this misunderstanding that there are an "unusual number of qualifications and disclaimers surrounding Freud's analysis of the oceanic feeling."[56] In a letter to Rolland describing *Civilization*, Freud cautions his friend, "Please don't expect from it any evaluation of the 'oceanic' feeling; I am experimenting only with an analytical version of it; I am clearing it out of the way, so to speak."[57] Considering how quick he had been to dismiss most forms of mysticism, Rolland's included, his phrasing is curious. The need to "clear something away" points to an obstacle or source of difficulty so what was it about the oceanic that troubled Freud?

While there is no clear or simple answer to this question, Colebrook reminds us that the "fantasy of an original and prehuman indifference" was tied not only to the oceanic but also to "the figure of the animal, the maternal [and] the 'savage.'"[58] This signifying chain is reproduced in *Civilization* as Freud identifies both the mysticism of "Yogis" and oceanic feeling with a regression to the primordial. In doing so, he effectively drew on the racial dynamics of *Totem and Taboo* to imply that non-white others are closer to the oceanic because they are less socially and culturally evolved. At the same time, Parsons contends that as a "man who embodied love extended to all men, who envisaged a civilization where all races and peoples could exist in harmony," Rolland symbolized for Freud the "as yet unextinguished hopes and aspirations of his youth—hopes that converged around the themes of assimilation, anti-Semitism, tolerance, and social justice."[59] This knot of complex and conflicting associations suggests that what Freud was at pains to clear away was not the oceanic per se but rather an aspect of his relationship to it—something that hindered the intellectual project of *Civilization* and perhaps of civilization. In the spirit of a speculative genealogy, I ask: Could that aspect be a sense of and desire for something revolutionary that was beyond his grasp but within reach of those marked by race?

References

Associated Press. "Denmark: Statue of Little Mermaid Vandalized Again." *The Washington Post*, July 3, 2020. https://www.washingtonpost.com/world/europe/denmark-statue-of-little-mermaid-vandalized-again/2020/07/03/4929f222-bd10-11ea-97c1-6cf116ffe26c_story.html.

Associated Press. "Denmark: Statue of Little Mermaid Vandalized Again." *AP News*, July 3, 2020. https://apnews.com/general-news-dc827daa685cc2ae262d7187da63dc3b.

Associated Press. "Rights Activists say Danes Unaware of Racism in Their Nation." *Voice of America*, July 3, 2020. https://www.voanews.com/a/6192168.html.

Bakker, Justine M. "Blue Black Ecstasy: Ellen Gallagher's Watery Ecstatic, Oceanic Feeling, and Mysticism in the Flesh." *Journal of the American Academy of Religion* 91, no. 2 (2023): 302–325.

Barounis, Cynthia. "Special Affects: Mermaids, Prosthetics, and the Disabling of Feminine Futurity." *WSQ: Women's Studies Quarterly* 44, no. 1 (2016): 188–204.

Bennett, Joshua. "Beyond the Vomiting Dark: Toward a Black Hydropoetics." In *Ecopoetics: Essays in the Field*, edited Angela Hume and Gillian Osborne. Iowa University Press, 2018.

Blaagaard, Bolette B. "Critical Memory and Anti-Racist Media: An Analysis of the Transnational Political Subjectivity of BLM-DK." *European Journal of Cultural and Political Sociology* 10, no. 4 (2023): 508–524.

Carlson, Kerstin Bree. "A Murder Trial in Denmark leaves a Nation Wondering What Constitutes a Hate Crime." *The Conversation*, December 7, 2020. https://theconversation.com/a-murder-trial-in-denmark-leaves-a-nation-wondering-what-constitutes-a-hate-crime-151446.

Colebrook, Claire. "The Counter-Oceanic Sea: Archive and Difference in *The Deep*." *Angelaki* 30, no. 1 (2025): 26–39.

Davis, Jalondra A. "Crossing Merfolk, the Human, and the Anthropocene in Nalo Hopkinson's *The New Moon's Arms* and Rivers Solomon's *The Deep*." *Journal of the Fantastic in the Arts* 32, no. 3 (112 (2021): 349–367.

"Denmark's Little Mermaid Vandalised." *France 24*, July 3, 2020. https://www.france24.com/en/20200703-denmark-s-little-mermaid-vandalised.

Er, Yanbing. "Underwater With the Feminist Waves: Black Gender and Oceanic Lifeworlds in Rivers Solomon's *The Deep*." *Diacritics* 51, no. 4 (2023): 60–77.

Erdbrink, Thomas. "A Black Man Was Tortured and Killed in Denmark: The Police Insist It Wasn't about Race." *New York Times*, June 30, 2020. https://www.nytimes.com/2020/06/30/ world/europe/denmark-bornholm-race.html.

Freud, Sigmund. "Letter from Sigmund Freud to Romain Rolland," July 20, 1929. In *The Letters of Sigmund Freud, 1873–1939*, ed. E.L. Freud, trans. T. Stern & J. Stern. Hogarth Press, 1961.

Gorelick, Nathan. "Extimate Revolt: Mesmerism, Haiti, and the Origin of Psychoanalysis." *CR: The New Centennial Review* 13, no. 3 (2013): 115–138.

Gumbs, Alexis Pauline. *Undrowned: Black Feminist Lessons from Marine Mammals*. AK Press, 2020.

Hartman, Saidiya. "Venus in Two Acts." *Small Axe* 12, no. 2 (2008): 1–14.

Henley, Jon. "Copenhagen's Little Mermaid Branded 'Racist Fish' in Graffiti Attack." *The Guardian*, July 3, 2020. https://www.theguardian.com/world/2020/jul/03/copenhagens-little-mermaid-branded-racist-fish-in-graffiti-attack.

Howard, Jonathan. "Swim Your Ground: Towards a Black and Blue Humanities." *Atlantic Studies* 20, no. 2 (2023): 308–330.

Mentz, Steve. *An Introduction to the Blue Humanities*. Routledge, 2023.

Murray, Adrienne. "Denmark Murder: Brothers Jailed over Bornholm Island Case." *BBC*, December 1, 2020. https://www.bbc.com/news/world-europe-55143044.

Nies, Betsy. "Transatlantic Mermaids: Literary and Cultural Fantasies from Copenhagen to Haiti and the United States." *Amaltea: Revista de Mitocrítica* 6 (2014): 305–327.

Noermark, Jakobs. "Murder Case: Prosecutor Wants to Disarm Racism Rumors – The Motive is a Personal Relationship." *Bornholms Tidende*, June 26, 2020. https://tidende.dk/bornholm/ drabssag-anklager-vil-afvaebne-racismerygter-motivet-er-en-personlig-relation/101278.

Parsons, William Barclay. *The Enigma of the Oceanic Feeling: Revisioning the Psychoanalytic Theory of Mysticism*. Oxford University Press, 1999.

Pressman, Jessica. "Hooked on Mermaids: Recuperating Personal Passion as Scholarly Research." *The Minnesota Review* 2023, no. 101 (2023): 113–122.

Reuters Staff. "Copenhagen's Little Mermaid labelled 'racist fish.'" *Reuters*, July 3, 2020. https://www.reuters.com/article/world/copenhagens-little-mermaid-labelled-racist-fish-idUSKBN24418I/.

Rolland, Romain. *Selected Letters of Romain Rolland*. Edited by Frances Dore and Marie-Laure Prevost. Oxford University Press, 1990.

Sexton, Jared. "Unbearable Blackness." *Cultural Critique* 90 (Spring 2015): 159–178.

Solomon, Rivers, Daveed Diggs, William Hutson, and Jonathan Snipes. *The Deep*. Simon and Schuster, 2019.

Spillers, Hortense. "Mama's Baby, Papa's Maybe: An American Grammar Book." *Diacritics* 17, no. 2 (1987): 65–81.

The Unknown Writer. Liner notes to *The Quest*, by Drexciya. Submerge. 1997.

VanHoose, Benjamin. "Denmark's Little Mermaid Statue Vandalized with 'Racist Fish' Graffiti." *People Magazine*, July 3, 2020. https://people.com/human-interest/little-mermaid-statue-in-denmark-vandalized-with-graffiti-that-reads-racist-fish/.

Notes

1. "Denmark's Little Mermaid Vandalised," *France 24*, July 3, 2020, https://www.france24.com/en/20200703-denmark-s-little-mermaid-vandalised.
2. See, for examples: "Denmark's Little Mermaid Vandalised"; Jon Henley, "Copenhagen's Little Mermaid Branded 'Racist Fish' in Graffiti Attack," *The Guardian*, July 3, 2020, https://www.theguardian.com/world/2020/jul/03/copenhagens-little-mermaid-branded-racist-fish-in-graffiti-attack; Associated Press, "Denmark: Statue of Little Mermaid Vandalized Again," *The Washington Post*, July 3, 2020, https://www.washingtonpost.com/world/europe/denmark-statue-of-little-mermaid-vandalized-again/2020/07/03/4929f222-bd10-11ea-97c1-6cf116ffe26c_story.html; Benjamin VanHoose, "Denmark's Little Mermaid Statue

Vandalized with 'Racist Fish' Graffiti," *People Magazine*, July 3, 2020, https://people.com/human-interest/little-mermaid-statue-in-denmark-vandalized-with-graffiti-that-reads-racist-fish/; and Reuters Staff, "Copenhagen's Little Mermaid Labelled 'Racist Fish,'" *Reuters*, July 3, 2020, https://www.reuters.com/article/world/copenhagens-little-mermaid-labelled-racist-fish-idUSKBN24418I/.

3. Jessica Pressman, "Hooked on Mermaids: Recuperating Personal Passion as Scholarly Research," *The Minnesota Review* 2023, no. 101 (2023): 119.
4. Henley, "Copenhagen's Little Mermaid."
5. Pressman, "Hooked," 120.
6. Adrienne Murray, "Denmark Murder: Brothers Jailed over Bornholm Island Case," *BBC*, December 1, 2020. https://www.bbc.com/news/world-europe-55143044.
7. Thomas Erdbrink, "A Black Man Was Tortured and Killed in Denmark: The Police Insist It Wasn't about Race," *New York Times*, June 30, 2020. https://www.nytimes.com/2020/06/30/world/europe/denmark-bornholm-race.html.
8. Kerstin Bree Carlson, "A Murder Trial in Denmark leaves a Nation Wondering What Constitutes a Hate Crime," *The Conversation*, December 7, 2020, https://theconversation.com/a-murder-trial-in-denmark-leaves-a-nation-wondering-what-constitutes-a-hate-crime-151446.
9. Bolette B. Blaagaard, "Critical Memory and Anti-Racist Media: An Analysis of the Transnational Political Subjectivity of BLM-DK," *European Journal of Cultural and Political Sociology* 10, no. 4 (2023): 508–524.
10. Erdbrink, "Black Man"; Jakobs Noermark, "Murder Case: Prosecutor Wants to Disarm Racism Rumors—The Motive is a Personal Relationship," *Bornholms Tidende*, June 26, 2020, https://tidende.dk/bornholm/drabssag-anklager-vil-afvaebne-racismerygter-motivet-er-en-personlig-relation/101278.
11. Henley, "Copenhagen's Little Mermaid"; Associated Press, "Denmark," *Washington Post*. The *Guardian* and *Washington Post* both draw from the Associated Press's reporting on the statue, and they describe Johansen's murder in the same way: "Rights activists on Thursday accused Danish officials of being unable to recognize racism after authorities said the killing of a biracial man by two white men was not racially motivated." No other international, American, or Danish news outlet appears to reference the vandalism in connection with Johansen. The sentence quoted above also opens a *Voice of America* article written by the Associated Press about the murder specifically. See Associated Press, "Denmark: Statue of Little Mermaid Vandalized Again," *AP News*, July 3, 2020, https://apnews.com/general-newsdc827daa685cc2ae262d7187da63dc3b; and Associated Press, "Rights Activists say Danes Unaware of Racism in Their Nation,"

Voice of America, July 3, 2020, https://www.voanews.com/a/6192168.html.

12. Henley, "Copenhagen's Little Mermaid."
13. Associated Press, "Rights Activists."
14. Jessica Pressman, ed., "Our Mermaid Craze," Special Issue, *ASAP Review* 18 (2024), https://asapjournal.com/cluster/our-mermaid-craze/.
15. Steve Mentz, *An Introduction to the Blue Humanities* (Routledge, 2023).
16. Justine M. Bakker, "Blue Black Ecstasy: Ellen Gallagher's Watery Ecstatic, Oceanic Feeling, and Mysticism in the Flesh," *Journal of the American Academy of Religion* 91, no. 2 (2023): 302–325; Jalondra A. Davis, "Crossing Merfolk, the Human, and the Anthropocene in Nalo Hopkinson's *The New Moon's Arms* and Rivers Solomon's *The Deep*," *Journal of the Fantastic in the Arts* 32, no. 3 (2021): 349–367; Alexis Pauline Gumbs, *Undrowned: Black Feminist Lessons from Marine Mammals* (AK Press, 2020); Jonathan Howard, "Swim Your Ground: Towards a Black and Blue Humanities," *Atlantic Studies* 20, no. 2 (2023): 308–330.
17. Joshua Bennett, "Beyond the Vomiting Dark: Toward a Black Hydropoetics," in *Ecopoetics: Essays in the Field*, eds. Angela Hume and Gillian Osborne (Iowa University Press, 2018), 103.
18. Alexis Pauline Gumbs, *Undrowned: Black Feminist Lessons from Marine Mammals* (AK Press, 2020), 7, 15.
19. Jonathan Howard, "Swim Your Ground: Towards a Black and Blue Humanities," *Atlantic Studies* 20, no. 2 (2023): 319.
20. Bennett, "Black Hydropoetics," 109.
21. Hortense Spillers, "Mama's Baby, Papa's Maybe: An American Grammar Book," *Diacritics* 17, no. 2 (1987): 72 (emphasis in the original).
22. Claire Colebrook, "The Counter-Oceanic Sea: Archive and Difference in The Deep," *Angelaki* 30, no. 1 (2025): 33.
23. Jared Sexton, "Unbearable Blackness," *Cultural Critique* 90 (Spring 2015): 172–173.
24. Colebrook, "Counter-Oceanic," 28.
25. Colebrook, "Counter-Oceanic," 28.
26. Saidiya Hartman, "Venus in Two Acts," *Small Axe* 12, no. 2 (2008): 11.
27. Hartman, "Venus," 11.
28. Cynthia Barounis, "Special Affects: Mermaids, Prosthetics, and the Disabling of Feminine Futurity," *WSQ: Women's Studies Quarterly* 44, no. 1 (2016): 190.
29. Betsy Nies, "Transatlantic Mermaids: Literary and Cultural Fantasies from Copenhagen to Haiti and the United States," *Amaltea: Revista de Mitocrítica* 6 (2014): 306.
30. The Unknown Writer, "Liner Notes," *The Quest* by Drexciya, Submerge, 1997, CD.
31. Unknown Writer, "Liner Notes."

32. Rivers Solomon, *The Deep* (Simon and Schuster, 2019), 20.
33. Solomon, *The Deep*, 28.
34. Solmon, *The Deep*, 28.
35. Yanbing Er, "Underwater With the Feminist Waves: Black Gender and Oceanic Lifeworlds in Rivers Solomon's *The Deep*," *Diacritics* 51, no. 4 (2023): 60–77.
36. Romain Rolland. *Selected Letters of Romain Rolland*, eds. Frances Dore and Marie-Laure Prevost (Oxford University Press, 1990), 86.
37. William Barclay Parsons, *The Enigma of the Oceanic Feeling: Revisioning the Psychoanalytic Theory of Mysticism* (Oxford University Press, 1999), 108.
38. Colebrook, "Counter-Oceanic," 28.
39. Colebrook, "Counter-Oceanic," 29.
40. Nathan Gorelick, "Extimate Revolt: Mesmerism, Haiti, and the Origin of Psychoanalysis," *CR: The New Centennial Review* 13, no. 3 (2013): 123.
41. Gorelick, "Extimate Revolt,"124.
42. Gorelick, "Extimate Revolt," 128.
43. Solomon, *The Deep*, 19.
44. Solomon, *The Deep*, 19–20.
45. Solomon, *The Deep*, 28.
46. Solomon, *The Deep*, 6.
47. Solomon, *The Deep*, 30.
48. Solomon, *The Deep*, 38.
49. Solomon, *The Deep*, 135.
50. Solomon, *The Deep*, 150.
51. Parsons, *Enigma*, 108.
52. Parsons, *Enigma*, 10.
53. Rolland, *Selected Letters*, 86–88.
54. Parsons, *Enigma*, 37–38.
55. Parsons, *Enigma*, 108.
56. Parsons, *Enigma*, 44.
57. Freud, "Letter from Sigmund Freud to Romain Rolland," 389, quoted in Ackerman, "Exploring Freud's Resistance,"10.
58. Colebrook, "Counter-Oceanic," 29.
59. Parsons, *Enigma*, 28.

Index

D. Leong, *Black Literature, Social Justice, and the Environment*,
https://doi.org/10.1007/978-3-032-19288-2

FSC
www.fsc.org
MIX
Papier | Fördert gute Waldnutzung
FSC® C083411

Kolibri 360 GmbH
Ferdinand-Jühlke-Straße 7
99095 Erfurt, Deutschland
produktsicherheit@kolibri360.de